ROOTS, TUB
PLANTAINS AN
in human

FOOD AND AGRICULTURE ORGANIZATION OF THE UNITED NATIONS
Rome, 1990

The designations employed and the presentation of material in this publication do not imply the expression of any opinion whatsoever on the part of the Food and Agriculture Organization of the United Nations concerning the legal status of any country, territory, city or area or of its authorities, or concerning the delimitation of its frontiers or boundaries.

David Lubin Memorial Library Cataloguing-in-Publication Data

FAO, Rome (Italy)

Roots, tubers, plantains and bananas in human nutrition.
(FAO Food and Nutrition Series, No. 24)

1. Roots 2. Tubers 3. Plantains 4. Bananas
5. Nutrition
I. Title II. Series.

FAO code: 86 AGRIS: S01
ISBN 92-5-102862-1

The copyright in this book is vested in the Food and Agriculture Organization of the United Nations. The book may not be reproduced, in whole or in part, by any method or process, without written permission from the copyright holder. Applications for such permission, with a statement of the purpose and extent of the reproduction, should be addressed to the Director, Publications Division, Food and Agriculture Organization of the United Nations, Via delle Terme di Caracalla, 00100 Rome, Italy.

© FAO, 1990

Printed in Italy

Contents

Tables

Figures

Acknowledgements

The book was prepared by Prof. O.L. Oke of the Abafemi Awolowo University, Ile-Ife, Nigeria and was edited and revised by Dr J. Redhead, Consultant, and Dr M.A. Hussain, Senior Officer, Community Nutrition Group, Nutrition Programmes Service, Food Policy and Nutrition Division. Valuable suggestions were offered by other staff members of the Nutrition Programmes Service and members of the FAO Interdepartmental Working Group on roots, tubers, plantains and bananas.

Preface

During the last 15 years the difficulties faced by many developing countries in satisfying their population's requirements with domestic food production have increased. Even with sustained efforts, it has not always been possible to meet the growing food demand by raising the domestic production of cereals. As a result widespread food shortages, hunger and malnutrition have persisted, particularly among the low-income groups in developing countries.

In order to improve the situation, the Member Governments of the Food and Agriculture Organization of the United Nations (FAO) at the 8th Session of the Committee on Agriculture (COAG) in 1985 recommended the adoption of measures to broaden the food base through the promotion of other local food crops of nutritional importance. More recently, at its 9th session, COAG further requested Member Governments to give high priority to production and consumption of roots, tubers, plantains and bananas in view of their important role in improving food security.

Although these crops have been for centuries the traditional staples in many developing countries they have until recently been relatively neglected by most national research institutes, extension services and by food supply planners. While part of this neglect can be attributed to difficulties in marketing and processing these perishable food crops, they have also suffered from a negative image as "poor people's food". Starchy roots and tubers, such as cassava, have been traditionally associated with poverty and accused of being a factor contributing to the development of kwashiorkor, a form of severe protein energy malnutrition. Since most of these food crops are consumed locally or sold in nearby small markets their actual contribution to the energy intake of rural populations producing them is not fully accounted for. Their consumption in urban areas is far from negligible, especially in Africa and in Asia. This is why it is time to bring out the positive attributes of these important foods and the increased contribution they can make to the nutritional welfare and food security of developing countries.

In this book the value of roots, tubers, plantains and bananas in human nutrition and their importance in human diet is reviewed. The purpose of this book is to promote their production and utilization as valuable components of a well-balanced diet, and to alleviate hunger and seasonal food shortages.

The book is intended for nutritionists, agriculturists, dieticians, community development workers, school teachers and economists. It is hoped that officials responsible for planning food supply, production and food imports and exports will find the facts presented useful for their work. Educationists will also find valuable information that will help them to promote changes in the food habits of population groups, particularly those who suffer chronically from energy deficiency and food insecurity.

P. Lunven
Director
Food Policy and Nutrition Division

1. Introduction

Roots and tubers belong to the class of foods that basically provide energy in the human diet in the form of carbohydrates. The terms refer to any growing plant that stores edible material in subterranean root, corm or tuber.

The development of root crops in the tropics was accelerated by the introduction of *gari*-processing technology into West Africa and by the promotion of cassava as a famine reserve by several colonial governments, such as the Dutch in Java and the British in West Africa and India. By 1880 the tapioca trade was well established in Malaya and by the turn of the twentieth century the production and trade of cassava products, especially starch, had been established by the Dutch in Java and by the French in Madagascar.

A further reason for the spread was the fact that during tribal warfare and invasions, the invader could not destroy or remove the food reserve, which could be kept conveniently under the ground, thus giving added food security to the population.

Historically, very little attention has been paid to root crops by policy-makers and researchers as most of their efforts have been concentrated on cash crops or the more familiar grains. Root crops were regarded as food mainly for the poor, and have played a very minor role in international trade. This misconception has lingered for so long because of the lack of appreciation of the number of people who depend on these root crops, and the number of lives that have been saved during famine or disasters by root crops.

It was cassava that saved the Rwanda-Burundi kingdoms in 1943 when potato blight destroyed all their production, and cassava also fed the Biafrans during the Biafran war in Nigeria in 1966-69.

As far back as 1844 Rev. John Graham has this to say about the potato:

"Oh! There's not in the wide world a race that can beat us,
From Canada's cold hills to sultry Japan,
While we fatten and feast on the smiling potatoes
Of Erin's green valleys so friendly to man."

There is an old saying of the Palananans of Micronesia where taro is the basis of the staple food that:

"the taro swamp is the mother of life" (Kahn, 1985).

The fact that these root crops are mainly starchy has led to the disparagement of their protein content, which is low compared to cereals. However, considering the quantities of root crops consumed a day, their protein contribution is often significant. In addition, root crops contain an appreciable amount of vitamins and minerals and may have a competitive production advantage in terms of energy yield per hectare over cereals produced in ecologically difficult conditions.

During the years 1980-87, the average rate of growth in food production (2.6 percent) in many countries of developing market economies, particularly Africa, has been either falling behind or barely keeping pace with the average annual rate of growth in population (about 3 percent), owing to land shortages and lack of foreign exchange to purchase agricultural inputs such as fertilizers, insecticides and machinery. Droughts, floods and other natural and induced calamities have contributed substantially to reduced food supplies. These food-deficit countries are now making substantial efforts to improve this situation, but their efforts are directed mainly toward improving production of staple cereals and increasing the importation of cereals by cash purchases or as food aid, thereby widening further the gap between local food production and food requirements.

In their present state under subsistence farming, the yield of many root crops is very low, but their genetic potential for producing increased yields is high and has not yet been fully exploited. In addition some root crops are highly adaptable, producing reasonable yields from marginal lands with highly erratic rainfalls. Crops such as cassava can serve as a valuable asset for household food security for subsistence populations in times of drought and under other unfavourable ecological conditions.

The purpose of this book is to review the value of roots and tubers in the human diet and to assess their contribution to the nutritional welfare and food security of people living in less developed countries. It is hoped that it will also help propagate knowledge about these crops and stimulate research for their genetic improvement with a view to increasing their production and utilization.

In its Food and Nutrition Papers and Series, the Food and Agriculture Organization of the United Nations (FAO) has already published books on five important food sources, namely *Rice and rice diets, Maize and maize diets, Milk and milk products in human nutrition, Wheat in human nutrition* and *Legumes in human nutrition.* The present study has been conceived on a similar pattern. It will summarize current knowledge about production, consumption, nutritive value, processing and cooking of roots and tubers and their contribution to human diets. The study includes all important root and tuber crops: cassava, yam, sweet potato, potato and aroids as well as two other starchy staples, banana and plantain. Although much of the research and development work on potato is done in temperate zones, the potato is included in the book because of its great potential for expansion into the tropics. Plantain and banana are also important starchy staples in many tropical countries.

In 1988 FAO published a study entitled *Root and tuber crops, plantains and bananas in developing countries:challenges and opportunities*, which provides a comprehensive review of the global production and consumption of these crops. Additional FAO studies in this field focus on the utilization and processing of roots, tubers, plantains and bananas. The present study provides a more detailed analysis of the role of these crops in human nutrition and is an essential supplement to the information provided in the earlier publications.

2. Origins and distribution

The suggested origins of root and tuber crops are illustrated in Table 2.1. These crops were dispersed by the Portuguese during their voyages for slaves, by both the Portuguese and Spanish in their missionary journeys, and by Arab traders. The genus *Dioscorea* (a variety of yam) has a wider diversity of origin with different species adapted to different ecosystems. *D. trifida* is indigenous to tropical America; *D. rotundata, D. cayenensis, D. bulbifera* and *D. dumetorum* are native to West Africa; *D. alata, D. esculenta* and *D. opposita* are indigenous to South Asia. *D. opposita* and *D. japonica* have their centre of origin in China.

Yams are the only root crops in which the Asian and African species developed independently of each other. Exchange of species was due to the influence of Portuguese explorers. They learned of the value of *D. alata* from the Indian and Malayan seafarers who used it on their ships on long voyages because it stored well and had antiscorbutic properties. The Portuguese soon adopted it and introduced it into Elmina and Sao Tome in West Africa. Subsequently, through the Atlantic slave trade, the Portuguese carried the African species *D. rotundata* and *D. cayenensis* and the Asian species *D. alata* to the Caribbean where they became important staple foods (Coursey, 1976). According to Coursey (1967), *D. alata* seems to have arisen from the wild relatives, *D. hamiltoni* and *D. persimilis* in the north and central parts of the southeast Asian peninsula, probably Burma or Assam. So also *D. esculenta* while *D. hispida, D. pentaphylla* and *D. bulbifera* origina ted from an Indo-Malayan centre. *D. rotundata* is of African origin, where it is known as "water yam", indicating that it was brought across the water or sea. *D. rotundata* is the most important African yam, especially in the forest zone, and is probably a hybrid of the other African yam, *D. cayenensis,* which is a savannah species. In West Africa it is grown in the roots and tubers belt, which extends 15°N and 15°S of the equator (Coursey, 1976; Okigbo, 1978; Nweke, 1981).

TABLE 2.1
Origins of tropical root crops

Root crop	Common name	Suggested origin
American species		
Ipomoea batatas	sweet potato	Tropical North America (Mexico, Central America and Caribbean)
Manihot esculenta	cassava, cocoyam	Tropical Central America (from Caribbean to Northeast Brazil)
Xanthosoma sagittifolium	new cocoyam, taro	Tropical Central America (from Caribbean to North Brazil)
Solanum tuberosum	potato	Andean South America (Colombia, Bolivia and Peru)
Dioscorea trifida	sweet yam	Tropical Central America (Guyana, Surinam)
African species		
Dioscorea rotundata	yam	Tropical West Africa
Dioscorea cayenensis	wild yam	Tropical West Africa
Dioscorea dumetorum	"	Tropical West Africa
Dioscorea bulbifera	"	Tropical West Africa
Asian species		
Dioscorea alata	yam	South Asia
Dioscorea esculenta	"	South Asia
Dioscorea opposita	"	South Asia
Colocasia esculenta	old cocoyam or taro	Southeast Asia
Musa acuminata	banana/plantain	Southeast Asia

Source: Adapted from Purseglove (1968, 1972).

Little is known about the origin of new world yams. They were of secondary importance in the pre-Colombian era. *D. trifida,* an Amerindian domesticate, appears to have originated on the borders of Brazil and Guyana, followed by a dispersion through the Caribbean (Ayensu and Coursey, 1972). Yams were taken to the Americas through precolonial Portuguese and Spanish expansion that began around 500 years ago. Historical records of *D. alata* in West Africa and of African yams in the Americas date back to the sixteenth century (Coursey, 1967).

Sweet potato, which originated in the Yucatan peninsula in Latin America, seems to be the most widely dispersed root crop. It is adaptable and can grow under many different ecological conditions. It has a shorter growth period than most other root crops (three to five months) and shows no marked seasonality: under suitable climatic conditions it can be grown all the year round. Adverse weather conditions rarely cause a complete crop loss. Hence sweet potatoes are planted as an "insurance crop", combined in mixed cropping with grains like rice in Southeast Asia, and with other root crops like cocoyam and yam in Oceania. It is a popular plant in the Philippines and in Japan because of its prostrate habit, which makes it resistant to damage by high winds such as hurricanes and typhoons (Wilson, 1977). Sweet potatoes have been cultivated since about 3 000 B.C. and were an important food for the Mayans in Central America and the Peruvians in the Andes. From ethno-historical records of Colombia, the reports of Spanish explorers and missionaries in Mexico and Peru, and of the Portuguese in Brazil, it is clear that sweet potato was common throughout the American tropics before 1492. The plant was further dispersed by Iberian and Portuguese explorers to the Pacific area in the sixteenth century. The Portuguese explorers subsequently transferred West Indian clones, grown in the western Mediterranean area, to Africa, India and the East Indies. Spanish traders also took sweet potatoes from Mexico to Manila. Later the sweet potato reached New Guinea and the eastern Pacific Islands and extended into China and Japan. It is now grown extensively in a wide range of environments between latitude 40°N and 40°S and from sea level up to an altitude of 2 300 metres.

The distribution of potato is also extensive. It originated in the high Andes of South America where it was adapted to the cold climate and short days prevailing in those latitudes. Wild cultivars are still found on elevated regions extending from the southwestern part of the United States of America to the southern part of South America, and more especially at high altitudes in Bolivia and Peru and in the coastal regions and nearby islands of southern Chile (Simmonds, 1976). When the original potato was first introduced to Europe it remained a botanical curiosity for more than a century and it did not flourish until a variety adapted to the longer day was evolved.

Spanish sailors introduced the potato to Spain as early as 1573. It was probably introduced into England by English seamen from captured Spanish ships around 1590. From Spain, potatoes spread throughout continental Europe; from England, they were dispersed throughout Great Britain to parts of northern Europe. By 1600, potatoes were sent from Spain to Italy and from there to Germany and in the same year they reached France.

Potatoes reached most other parts of the world through European colonial activities. North America received potatoes from England in 1621, British missionaries took potatoes to Asia in the seventeenth century and Belgium missionaries carried them to the Congo in the nineteenth century. The potato was brought to India in the sixteenth century by Portuguese traders and within about 200 years it had spread all over India. It was taken to Bhutan, Nepal and Sikkin from India. In Africa, introduction of potatoes followed colonization. Possibly, its antiscorbutic properties persuaded seafarers to stock it in their ship's store and to encourage people to grow it wherever they visited.

Like sweet potato, potato also exhibits a growing period of about four months, shorter than many other root crops. The indigenous South American cultivars will develop tubers at longer day lengths than other root crops and numerous cultivars will even tolerate the extreme day length of 24 hours of the polar summer (Kay, 1973). So the spread was very easy.

A typical example of a root crop that can tolerate drought and poor husbandry is cassava. Cassava originated in tropical America but the precise area of its origin is not known. The two probable areas suggested are the Mexican and Central American area or northern South America. It was first introduced into the Congo basin as early as 1558 by the Portuguese. It then spread rapidly through Angola, Zaire, Congo and Gabon and later to West Africa. There was a separate introduction to the east coast of Africa and to Madagascar in the eighteenth century by Portuguese and Arab traders, after which it rapidly became a dietary staple throughout many lowland tropical areas (Jones, 1959). The cultivation of cassava in Africa increased during the nineteenth and twentieth centuries as a result of encouragement by administrative authorities, who recognized its value as a famine relief crop. According to Kahn (1985), after the First World War, the farmers of Ruanda-Urundi, now the independent nations of Rwanda and Burundi, at first refused to take advice from the Belgians to plant cassava, because they had enough potatoes. But in 1924 the Belgians issued a strict order to grow cassava and recruited 60 000 porters to carry 5 000 tonnes of cassava stakes around the region for planting, so the farmers finally accepted.

Cassava was taken to India by the Portuguese in the seventeenth century. In about 1850, it was transported directly from Brazil to Java, Singapore and Malaya. Cassava was introduced to the South Pacific territories during the first half of the nineteenth century by missionaries and travellers but its importance varies from island to island. At present, cassava is grown throughout tropical and subtropical areas approximating 30°N and 30°S of the equator and up to an altitude of 1 500 metres.

The spread of root crops was facilitated by their ability to thrive under varied tropical conditions. Their level of water tolerance varies considerably, ranging from the waterlogged conditions required for taro to the drought tolerance and minimal water supplies needed for cassava once it has been established (Wilson, 1977). It was the requirement of flooded conditions for taro, *Colocasia esculenta,* that convinced anthropologists that these yams were the first irrigated crops, and that the ancient "rice" terraces of Asia were originally constructed for them (Plucknett *et al.*, 1970). Tannia *(Xanthosoma sagittifolium)* on the other hand cannot tolerate being waterlogged (Onwueme, 1978).

Xanthosoma, or new cocoyam, had its origin in South America and the Caribbean. The Spanish and Portuguese introduced it to Europe and were also responsible for spreading it to Asia. It moved from the Caribbean in the late nineteenth century, first to Sierra Leone and then to Ghana. In West Africa, *Xanthosoma* is more important than *Colocasia,* being popular for its corm, cormels, leaves and young stems. Although *Xanthosoma* is relatively new to the Pacific region, it has spread rapidly and widely, becoming quite an important crop in many of the islands. It is also widely cultivated in Puerto Rico, the Dominican Republic and Cuba and is important along the coastal mountains of South America, in the Amazon basin and in Central America.

Colocasia originated in India and Southeast Asia. About 2 000 years ago it spread to Egypt and thence to Europe (Plucknett *et al.,* 1970). Subsequently it was taken from Spain to tropical America and then to West Africa. It was used in feeding slaves and was transferred to the West Indies with the slave trade (Coursey, 1968). In order to distinguish it from the newer species, *Xanthosoma, Colocasia* was referred to as "old yam" in West Africa whereas *Xanthosoma* is called "new yam". *Colocasia* is a staple food in many islands of the South Pacific, such as Tonga and Western Samoa, and in Papua New Guinea. *Colocasia* and *Xanthosoma* will tolerate shade conditions and so they are often planted under permanent plantations like banana, coconut, citrus, oil palm and especially cocoa. Therefore they are sometimes collectively referred to as cocoyams.

The banana is believed to have originated in Southeast Asia, having been cultivated in South India around 500 BC. From here it was distributed to Malaya through Madagascar and then moved eastward across the Pacific to Japan and Samoa in the mid-Pacific at about AD 1000. It was probably introduced to East Africa around AD 500 and had become well established in West Africa by AD 1400. It finally arrived in the Caribbean and Latin America soon after AD 1500 (Simmonds, 1962; 1966; 1976). By the end of the eleventh century, banana had spread widely throughout the tropics. In South America it was found as far south as Bolivia and was cultivated in most of Brazil. In Africa banana growing extended from the Sahara to Tanzania in the east and from Côte d'Ivoire through the Congo to Zaire in the west and central areas.

TABLE 2.2
Minor root crops of local importance

Local names	Root crop species	Suggested origin	Other names
Chayote	*Sechium edule*	Mexico	Chinchayote, Guisquil (Spanish)
Jicama	*Pachyrhyzus*	Mexico	
Yambean	*Pachyrhyzus and Sphenostylis stenocarpa*		
Arrow root	*Maranta arundinacea*	Polynesia	*Pana, Panapen*
Arracachia	*Arracacia xanthorrhiza*		
Oca	*Oxalis tuberosa*		
Queensland arrowroot	*Cana edulis*		
Topee Tambo	*Calathea allouia*		
Ulluco	*Ullucus tuberosus*	Mellocco, oca-quira	
Yacon	*Polymnia sonchifolia*		

Apart from the major root crops discussed in this book, other root crops exist in different parts of the world, mainly in the Andean region, and are of local importance. Some of these are shown in Table 2.2.

3. Production and consumption

PRODUCTION

According to a recent FAO estimate, virtually every country in the world grows some species of root crop. Most of the root crops considered in this study require tropical conditions and are restricted to Africa, Asia and Latin America. Only potatoes and some varieties of sweet potato are grown in large quantities in the temperate zone. These root crops are often the main dietary staple for low-income consumers. They are grown by farmers as subsistence crops on small plots of land ranging from two to 20 hectares depending on the region.

It has been estimated that about 82 percent of Paraguayan farmers grow cassava as a subsistence crop on small holdings, and whenever they move on to virgin land they first plant cassava. In Latin America, 75 percent of the cassava farms are about 20 hectares or less, whereas in Java and Kerala the holding is about two hectares. In Thailand most producers have less than one hectare devoted to cassava. In 1982/83 the cultivation and harvesting of some 19 million tonnes of cassava in Thailand was carried out entirely by approximately 1.2 million smallholders who obtained yields of between 13 and 15 tonncs per hectare (FAO, 1984). Most of this production was processed, with 85 percent transformed into chips and pellets for animal feed and 15 percent used for starch production. Very little was used directly for human consumption.

In Africa these root crops are usually subsistence crops grown mainly as food, so the farmer keeps sufficient to feed his family and sells only the surplus. However, there is now a growing commercial market for them. Cassava is commercially processed into *gari,* a staple food in parts of Nigeria, and into *kokonte* in Ghana. In Brazil about 70 percent of the cassava harvest is marketed (Lynam and Pachico, 1982).

In subsistence production of cassava, yields are often low as a result of poor cultivation practices. Cassava is often grown on marginal land and as it grows relatively well on poor soils, with limited inputs, it is often planted as the last crop in a shifting cultivation system. On average, weeds reduce cassava yield by 59 percent. On newly cleared land no positive yield response is observed to either nitrogen or potash fertilizers. On poor land there may be some response to nitrogen, but fertilizer is not extensively applied. Even in Java in Indonesia, where land use is very intensive and the cost of fertilizer is heavily subsidized, only 8.1 kg of fertilizer/ha was used for cassava compared to 178.9 kg/ha average for all the other crops. In Brazil only about 9 percent of the cassava area is fertilized.

Most research to improve root crop production has been devoted to potato in temperate countries and in the tropics, especially at the International Centre for Potato Research (CIP) in Peru, and so it is not surprising that potato yields are much higher than those of the other root crops. In some parts of Latin America, however, it is still grown by subsistence farmers on a small scale, as part of a complex multiple-cropping system, on one to two hectare plots with low yields. In temperate zones and in cool highland areas, where it is usually grown under irrigation conditions and as a sole crop, yields are often very high. It is produced only in limited quantities in the tropics, cassava and sweet potatoes being the major crops there.

In 1982, the CIP estimated that there has been an overall increase in the total production of root crops in developing countries during the years from 1961 to 1979. However, when considered individually and regionally, production of crops like cassava has gone up, sweet potatoes have remained stagnant, while production of potatoes has decreased in industrialized countries, but increased in developing countries. Per caput production of root crops had fallen during this period in most developing countries. In sub-Saharan Africa, output of root crops, except sweet potato, failed to keep pace with population growth. In Latin America and the Caribbean, since 1970, production trends of starchy staples as a group have been negative (FAO, 1988a). Various reasons have been adduced for the decrease in production, including infestation by insects, parasites and diseases, bad weather and marketing problems.

That part of the root and tuber crop harvest which is produced by subsistence farmers for their own consumption does not enter the commercial marketing channels. It is thus difficult to obtain accurate data on total production of these crops. Today, FAO's statistics are the best available guide to global production of these crops.

Tables 3.1 and 3.2 provide production, acreage and yield figures for roots and tubers in various regions of the world. Among the five root crops listed, potatoes occupy a land area of about 20 million hectares or 44.3 percent of the total area of 46 million hectares devoted to world production of root crops. The potato is increasingly important in developing countries and is a good source of nutrients in the diet. Its protein:calorie ratio is as high as that of wheat (Table 4.10) and its productivity in terms of energy and protein per hectare per day is greater than that of most other staple food crops (Table 4.1).

Potato has the highest percentage of world production, accounting for 52.9 percent of the total in 1984, followed by cassava covering 14 million hectares (21.9 percent) and accounting for 30.9 percent of total production; then sweet potatoes covering about eight million hectares (16.9 percent) and accounting for 19.9 percent of total production. Yams cover about 3 million hectares (5.5 percent) with 4.3 percent of total production and the least important, taro, occupies 1 million hectares (2.5 percent) with 1.0 percent of total production.

Table 3.1 shows that potato is growing over a wide geographical spread in the countries producing this root crop, but the leading producers are all in temperate-zone areas (Table 3.3). Of the total of 130 potato-producing countries, 95 are developing countries and in 1978-81 they were able to produce less than 10 percent of the world production. However, by 1985 developing countries accounted for about one-third of world production, with China producing 60 percent of this contribution. The increase has been particularly significant in the Near East where production had gone up by 130 percent, in the Far East by 180 percent and in Africa by 120 percent. Potato is also a potentially high-yielding crop. The present average yield is only about 10 t/ha in developing countries but yields as high as 72 t/ha have been recorded on experimental research plots in the Netherlands and this

TABLE 3.1

World area, production and yield of root crops in 1984

	Cassava			Sweet potatoes			Yams			Taro			Potatoes		
	10^3 ha	10^3 t	t/ha	10^3 ha	10^3 t	t/ha	10^3 ha	10^3 t	t/ha	10^3 ha	10^3 t	t/ha	10^3 ha	10^3 t	t/ha
Africa	7 482	51 002	6.8	841	5 136	6.1	2 395	24 426	10.2	926	3392	3.7	660	5830	8.8
North and Central America	165	922	5.6	217	1 446	6.7	49	317	6.4	2	20	9.6	746	20 353	27.3
South America	2 311	26 861	11.6	162	1 431	8.9	40	334	8.4	-	-	-	926	10 355	11.2
Asia	4 171	50 000	12.0	6 390	108 634	17.0	16	166	10.2	179	2 027	11.3	5 814	82 135	14.1
Oceania	22	235	10.7	116	559	4.8	18	246	13.7	46	319	6.9	48	1 130	23.5
Europe and USSR	-	-	-	13	131	10.4	-	-	-	-	-	-	12 109	192 406	16.4
World	14 151	129 020	9.1	7 739	117 337	15.2	2 518	25 489	10.1	1 153	5 758	5.0	20 303	312 209	15.4

Source: Chandra, 1988.

TABLE 3.2

World area and production of root crops in 1984 (in percentages)

	Cassava		Sweet potato		Yam		Taro		Potato		Total	
	Area	Prod.	Area	Prod.	Area	Prod.	Area	Prod.	Area	Prod.	Area	Prod.
Africa	52.9	39.5	10.9	4.4	95.1	95.8	80.3	59.0	3.2	1.9	26.8	15.2
North and Central America	1.2	0.7	2.8	1.2	2.0	1.2	0.2	0.3	3.7	6.5	2.6	3.9
South America	16.3	20.1	2.1	1.2	1.6	1.3	-	-	4.6	3.3	7.5	6.6
Asia	29.5	38.8	81.3	92.6	0.6	0.7	15.5	-	28.6	26.3	36.0	41.2
Oceania	0.2	0.2	1.5	0.5	0.6	1.0	4.0	5.5	0.2	0.4	0.5	0.4
Europe and USSR	-	-	0.2	0.1	-	-	-	-	59.6	61.6	26.4	32.6
World	30.9	21.9	16.9	19.9	5.5	4.3	2.5	1.0	44.3	52.9	100	100
Total developing countries	98.7	99.1	94.2	98.2	97.3	97.8	95.8	94.1	36.5	31.5	70.2	63.0
Total developed countries	1.3	0.9	5.8	1.8	2.7	2.2	4.2	5.9	63.5	68.5	29.8	37.0

Source: FAO, 1985.

TABLE 3.3

Leading root crop producers in 1984 (percentage of total)

Cassava		Sweet potatoes		Yams		Taro		Potatoes	
Brazil	16.4	China	83.4	Nigeria	72.6	Nigeria	30.5	USSR	27.3
Thailand	15.4	Indonesia	1.7	Côte d'Ivoire	9.2	China	25.4	China	17.6
Zaire	11.4	Viet Nam	1.6	Ghana	3.4	Ghana	12.7	Poland	12.0
Indonesia	10.9	India	1.4	Benin	2.7	Japan	6.9	USA	5.3
Nigeria	9.1	Japan	1.2	Togo	1.8	Côte d'Ivoire	5.4	India	3.9
India	4.4	Philippines	0.9	Cameroon	1.6	Papua N. G.	3.2	Germany DR	2.6
Tanzania	4.3	Rwanda	0.8	Ethiopia	0.8	Philippines	2.6	Germany FR	2.4
China	3.2	Korea Rep.	0.8	Cent. A. Rep.	0.8	Burundi	1.7	UK	2.3
Mozambique	2.4	Uganda	0.7	Zaire	0.7	Egypt	1.7	Netherlands	2.1
Viet Nam	2.2	Bangladesh	0.7	Brazil	0.7	Madagascar	1.6	Romania	2.1
Total	79.7	Total	93.2	Total	94.3	Total	91.7	Total	77.6
Total no. of countries	95		106		43		29		130
Developing	95		100		41		28		95
Developed	0		6		2		1		35

Source: Adapted from Chandra, 1988.

could be further increased by the use of improved varieties under conditions of good husbandry (Doku, 1984). At present the normal recorded yield for the United States of America is about 27.3 t/ha.

In spite of the low production, potato has become an acceptable foodstuff in a number of developing countries, including China, Bolivia, Colombia, Ecuador, India, Guatemala, Kenya and Rwanda (see Table 3.4). After China, India is the leading producer, accounting for 3.6 percent of world production, followed by Turkey (1.1 percent), Brazil and Colombia (0.8 percent). Together these four countries account for over 50 percent of the production in developing countries but only 7 percent of the world production.

Present research projects at CIP include breeding varieties to tolerate tropical temperatures at altitudes down to 1 000 ft (300 m) or even lower. Great progress has recently been made in the field of tissue culture and genetics and it may not be long before the potato also becomes a common tropical root crop. This will materially assist the provision of additional food for the ever increasing population in this part of the globe. At present potatoes provide only a small percentage of dietary calories in most developing countries, as indicated in Table 3.4. Cassava and sweet potatoes are more important root crops providing a range from 57.9 percent of the calories in Zaire to 35.2 percent in Angola.

Over the last 20 years (1965-84), cassava production worldwide has increased by over 330 percent. This corresponds to an annual growth rate of 4.3 percent which is substantial for any food crop (Chandra, 1988). Calculated recent changes in world production in 1986 with reference to base year 1984 showed that production of cassava has increased by 5.2 percent, yam by 4.8 percent and taro by 3.7 percent. While world production (including temperate-zone production of sweet potato and potato) has decreased by 3.8 percent and 1.8 percent respectively, the position of these two crops in some developing countries continues to strengthen. Production of sweet potato between 1969-71 and 1981-83 grew by 3.4 percent per annum in sub-Saharan Africa (FAO, 1986a) and by percentages varying from 6.3 percent per annum (Viet Nam) to 1.1 percent per annum (Thailand) in selected countries of Asia (FAO, 1987b). The increase in potato production for some Asian countries was 7.8 percent per annum for India, 6.2 percent

TABLE 3.4

Ten developing market economies with highest calorie intake derived from root crops

(in percentages)

Potatoes		Sweet potatoes		Cassava		Yam		Cocoyam		All roots	
Bolivia	9.3	Tonga	39.4	Zaire	55.7	Côte d'Ivoire	18.1	Samoa	15.9	Zaire	57.9
Peru	6.7	Solomon Is.	30.2	Congo	50.6	Togo	15.5	Ghana	11.4	Congo	54.2
Ecuador	5.6	Burundi	20.3	Cent. Afr. Rep.	38.6	Nigeria	15.5	Solomon Is.	6.8	Centr. Afr. Rep.	49.4
Colombia	3.8	P. New Guinea	15.6	Mozambique	36.1	Benin	12.0	P. New Guinea	4.5	Tongo	45.7
Chile	3.3	Rwanda	15.5	Angola	31.9	C. Afr. Rep.	8.5	Fiji	3.7	Solomon Is.	44.2
Turkey	3.2	Martinique	5.0	Comoros	25.2	Solomon Is.	7.2	Gabon	3.1	Burundi	38.2
Argentina	3.0	Comoros	4.6	Tanzania	24.1	Gabon	5.4	Côte d'Ivoire	2.3	Mozambique	37.1
Burundi	2.7	Uganda	4.5	Liberia	21.0	Jamaica	5.3	Centr. Afr. Rep.	2.3	Ghana	35.5
Rwanda	2.4	Guadeloupe	3.1	Benin	20.4	Ghana	5.3	Nigeria	2.2	P. New Guinea	35.2
N. Caledonia	2.3	Madagascar	3.0	Togo	19.8	P. N. Guinea	5.0	Togo	1.0	Angola	35.2

Source: Horton *et al.*, 1984.

for China, 10.2 percent for Sri Lanka and 13.8 percent for Viet Nam, between 1970-72 and 1982-84. Since 1970 there were also significant gains in potato production in Cuba, Colombia, Venezuela and most of Central America, owing to the adoption of new technologies.

Nigeria is the highest producer of yams, producing about 73 percent of the world total, most of which is used locally as food. Other leading producers are the West African countries of Côte d'Ivoire (9.2 percent), Ghana (3.4 percent), Benin (2.7 percent), Togo (1.8 percent) and Cameroon (1.6 percent). Virtually all the world production of yam is from West Africa, with *D. rotundata* the most important variety and *D. cayenensis* the least important. The other developing countries that produce some yam are in Central and South America, e.g. Haiti, Chile and Ecuador.

The Samoan islanders of the South Pacific region derive nearly 16 percent of their calorie intake from the consumption of cocoyam (aroids), but these root crops are less important in Africa. In Ghana they contribute about 11 percent of calories to the diet but in Nigeria and Côte d'Ivoire their contribution is only about two percent of calorie intake (see Table 3.4.). In Latin America and the Caribbean the production of aroids rose at less than one percent per year between 1969-71 and 1982-84 and did not keep pace with population growth. Similarly in Oceania annual growth in aroid production was sluggish at 1.3 percent and in South and Southeast Asia growth in production during recent years has been negligible.

Mention must be made of banana and plantain which have made a significant contribution to some subsistence economies, especially as the labour cost is relatively lower even than for that of cassava. Labour requirements for the production of various root crops in Nigeria, for example, are shown in Table 3.5.

Plantains and cooking bananas are grown and utilized as a starchy staple mainly in Africa, where production in 1985 was almost 17 million metric tonnes out of total developing-country production of 24 million metric tonnes. Of this total, South American production contributed about 4 million tonnes with the remainder produced in Asia, Central America and Oceania. In most of these regions annual production growth rates between 1969-71 and 1982-84 were only 1.7 to 1.8 percent, and well below the rate

TABLE 3.5
Labour requirements for production of various staple crops in Nigeria

Crop	Working days/ha	Working days/mt	Working days/mcal
Yam	325	45	69.31
Cassava	183	21	20.57
Maize	90	121	35.51
Rice	215	145	59.92

Source: Nweke, 1981.

of population increase. Although overall production in South and Southeast Asia remained low, the annual rate of growth in production was more encouraging at 4.9 percent total for both plantain and banana.

Cultivation of *ensete, Ensete ventricosum,* is limited to Ethiopia, where it is a staple food crop of the people of the southern highlands. It resembles a banana plant and is often called the "false banana". It does not produce an edible fruit and is harvested as a food source before flowering. The starchy portions of the swollen pseudo-stem and the underground corm are edible. It has been estimated that between 7 to 8 million people in the south and southwestern part of Ethiopia depend on fermented starchy staples prepared from *ensete* (FAO, 1985b).

CONSUMPTION

Root crops make an important contribution to the diet of many people in tropical countries, being consumed, as in the case of cassava, as a basic source of low-cost calories, or as a supplement to cereals. The cost of calories from cassava is about 25 to 50 percent that of the locally produced traditional grains and pulses (Goering, 1979), but other root crops, such as yams, are considerably more expensive. In most developing countries the dietary staples consist of starchy foods, which usually include some root crops. As indicated in Table 3.6, tropical roots may supply from as much as 1 060 calories per caput per day, (56 percent of the total daily calorie intake in Zaire), to as little as 200 calories (eight percent of the total daily intake in Belize).

Root crops are consumed not only by adults but are also important items in the diets of children. For example in Ghana and Nigeria infants are often weaned to an adult diet consisting of cassava or plantain. In Zaire, cassava *fufu* is the second most popular solid food for children under the age of one year, while in Cameroon, cassava is commonly given to infants 6 to 11 months old. Cassava can only form the basis for an adequate diet if it is consumed with other protein-rich foods such as oilseeds, pulses and fish. Young children have a limited stomach capacity and are unable to eat enough bulky foods, such as roots and tubers, to meet their energy needs. The results of a recent survey provide useful data on the frequency of cassava consumption in Zaire. In the survey area, sweet cassava is eaten raw in some localities as a snack or is boiled as *ebe*. The bitter varieties of cassava are made into *fuku,* a cassava gruel, to which various amounts of corn are added, depending on the season. *Mpondu,* a vegetable dish made from cassava leaves, is frequently eaten with the *fuku*. In many areas, *fuku* with *mpondu* or other cassava derivatives was eaten about twice a day by over 90 percent of the population in the 24 hours before the interview.

The income elasticity of demand for root crops is low but positive, and the cross elasticities of demand among cereals and root crops are high so substitution is not difficult. A national socio-economic survey conducted in Indonesia in 1980 showed that the per caput consumption of fresh cassava tends to increase as minimal income level increases, but stabilizes or decreases at the higher income levels.

A similar result has been observed in Brazil where the elasticity of demand for cassava is positive at low income levels, and in Ghana where there is no further tendency for the consumption to increase as per caput income increases to levels well above subsistence. In Indonesia, there is a high cross elasticity of demand between cassava and rice. If better production or storage techniques could result in reduced consumer costs of cassava products, then the prospects for an increased consumption of cassava would improve appreciably (Cock, 1985). With some other root crops, especially yams, the consumption tends to increase with rising income as yams are a

TABLE 3.6

Tropical root crops as a source of calories in selected countries, 1974

	Population (mid-1975, millions)	GNP unit per caput (market prices, 1975)	Average total calorie consumption[1] (per caput per day)	Calories from root crops[1]	Percentage of total calories from root crops
Zaire	24.7	140	1 880	1 060	56
Ghana	9.9	590	2 320	870	38
Togo	2.2	250	2 220	850	38
Côte d'Ivoire	6.7	540	2 650	820	31
Nigeria	75.0	340	2 080	540	27
Cameroon	7.4	280	2 370	530	22
Paraguay	2.6	580	2 720	450	17
Madagascar	8.8	200	2 390	370	15
Bolivia	5.6	360	1 850	290	16
Guinea	5.5	130	2 000	290	14
Uganda	11.6	230	2 100	300	14
Peru	15.4	760	2 330	310	13
China (PRC)	822.8	380	2 360	270	12
Indonesia	132.1	220	2 130	250	12
Kenya	13.4	220	2 120	200	9
Brazil	107.0	1030	2 520	230	9
Jamaica	2.0	1110	2 660	230	9
Belize	0.1	670	2 440	200	8

Notes:
[1]Figures rounded to the nearest 10.
Calorie consumption data are from FAO. Population and income figures are from *World Bank Atlas, 1977*.
Source: Goering, 1979.

relatively expensive food. In some places there may also be a strong cultural preference for particular foods, such as sweet potatoes. However, the general tendency is that cereals are preferred to root crops, while wheat and rice are preferred to the coarser grains.

As shown in Table 3.7 root crops contribute about 78 percent of the total calorie intake in the Group I region of sub-Saharan Africa, which is mainly in the tropical rain forest belt, and about 43 percent of the total calories in the Group II area, whereas in the more arid zones in Group III, cereals are more prominent. The details of the FAO country classification by groupings for sub-Saharan Africa are as follows:

Group I: Central African Republic, Congo, Mozambique, Zaire. In these countries both production and consumption patterns are dominated by cassava, which accounts for over 50 percent of staple food consumption. Cereals provide 30 percent, and nearly one-third of these are imported.

Group II: Angola, Benin, Burundi, Cameroon, Comoros, Equatorial Guinea, Gabon, Ghana, Côte d'Ivoire, Nigeria, Rwanda, Tanzania, Togo, Uganda. A far more diverse production/consumption pattern is prevalent in this group. While roots and plantains are the main staple foods, cassava consumption is much less important than in the previous group. Included here are countries typical of the West African yam-producing belt. While plantains, sweet potatoes and taro are important foods in individual countries cereals account for one-half of calories consumed. Approximately 30 percent of cereals are imported.

Group III: Botswana, Burkina Faso, Cape Verde, Chad, Ethiopia, Gambia, Guinea, Guinea-Bissau, Kenya, Lesotho, Liberia, Madagascar, Malawi, Mali, Mauritania, Mauritius, Namibia, the Niger, Reunion, Sao Tome and Principe, Senegal, Seychelles, Sierra Leone, Somalia, Sudan, Swaziland, Zambia, Zimbabwe. In these countries cereals play a far greater role in both production and consumption, although roots are often staple foods in particular regions. The proportion of cereal consumption met from imports is also generally lower, averaging under one-fifth of the total.

TABLE 3.7

Levels of consumption of staple foods in sub-Saharan Africa, 1981-83

	Group I[1]	Group II[1] (kg per caput/per annum)	Group III[1]	Total
Starchy staples	453.4	274.0	45.1	205.1
Cassava	407.4	123.0	21.3	117.8
Yams	6.6	72.4	3.5	36.8
Sweet potatoes	6.6	20.3	5.0	12.5
Plantains	26.2	39.1	2.0	22.7
Others	6.6	19.2	13.3	15.3
Cereals	39.7	83.8	134.1	98.3
		(Percentage, in calorie equivalent)		
Starchy staples	78	49	9	39
Cassava	70	22	4	24
Yams	1	14	1	7
Sweet potatoes	2	3	1	2
Plantains	4	6	-	4
Others	1	4	3	2
Cereals	22	51	91	61

[1]See text for explanation of groups.
Source: FAO, 1987.

In the Pacific root crops still supply from 15 to 43 percent of the dietary energy, the type depending on the island: taro and yam provide 43 percent of the energy in Tonga while sweet potato, taro and yam are the chief suppliers of energy in Papua New Guinea and the Solomon Islands. The difference in cereal and root crops intake between rural and urban areas is striking. The rural population consumes as much as twice the quantity of root crops but less than a tenth of the quantity of cereals than the urban population. This is related to the high cost of transportation and the short shelf-life of fresh root crops. The picture is similar in Latin America and the Caribbean (Table 3.8). In these areas the cost of production of root crops is

TABLE 3.8

Rural/urban consumption of root crops in Latin America and the Carribean, selected crops and countries

	Rural consumption (kg/head/year)	Urban consumption (kg/head/year)
Fresh cassava		
Brazil (1975)	11.2	2.7
Paraguay (1976)	180	35
Colombia (1981)	25.5	8.3
Cuba (1976)	30.0	12.4
Farinha da mandioca		
Brazil (1975)	29.4	9.7
Potato		
Peru (1981)	110	45
Yam		
Colombia (1981)	5.9	2.8

Sources: Lynam, J.K. and Pachico, D., *Fresh cassava in Brazil, Cuba and Paraguay, farinha de mandioca*, 1982.
Sanint, L.R. *et al.*, *Fresh cassava and yam in Colombia*, 1985.
Scott, G., *Potato in Peru*, 1985.

so high compared to that of cereals that some traditional root crops have become a luxury, except for potato in Bolivia and Peru and cassava in Brazil and Paraguay. In the Caribbean, cereals are definitely more important than roots in the diet though crops like plantain still contribute a substantial part of the dietary energy. In addition, some of the root crops produced are used for animal feed. About 33 percent of the cassava and about three to four percent of the other root crops are used for this purpose.

Oñate *et al.* (1976) have shown in Table 3.9 that the consumption of root crops in Southeast Asia ranges from 6 kg/caput/year (16 g/caput/day) in Cambodia to 113 kg/caput/year (310 g/caput/day) in Indonesia. The per caput consumption of potato in Singapore is very high (9 kg/year or 25 g/day) compared to the other countries in the region (0 to 7 g/day). Banana makes up a substantial proportion of the Filipino food intake, ranging from 15 g/caput/day in eastern Visayas to 40 g in western Visayas region.

Tables 3.10 and 3.11 show the results of dietary surveys in eight regions in the Philippines varying from the urban population in metropolitan Manila to the rural populations in lowland and mountain areas.

Daily nutrient and food allowances for each individual were obtained from tables prepared by the Food and Nutrition Research Institute. Nutrients in the foods were calculated from food composition tables. Allowances for all members of the household were added together and divided by the number of household members to get the per caput allowances per household. The daily per caput nutrient allowances for each region were obtained by dividing the sum of allowances for all households surveyed in the area by the total number of individuals. Thus the percentages quoted in Table 3.10 give a comparative regional survey of dietary adequacy. The population with the poorest diet in terms of calorie, protein and iron intake is that of eastern Visayas where starchy roots and tubers contribute the bulk of the calories. Table 3.11 shows that the intake of supplementary foods such as legumes, fruit and vegetables, milk, eggs and oils and fats is exceptionally low in this region.

This reinforces the argument that an increased intake of dietary calories from roots and tubers must be supplemented by a variety of other foodstuffs in order to achieve a balanced diet. Studies by Oñate *et al.*, 1976, show that in the Philippines the daily consumption of root crops varies with the region, ranging from about eight grams in metropolitan Manila to 222 grams in the rural area of eastern Visayas. This is true of many countries where food consumption in urban areas is characterized by a decrease in the consumption of root crops and an increase in the consumption of convenience foods from cereals and animal protein (see Table 3.11). The situation is different for processed root products with a relatively high bulk density and an extended shelf-life. In Ghana for example, after fresh cassava is made into dried *gari,* it can be transported relatively cheaply to the urban areas where it is as popular as in the rural areas. In the case of yam, surprisingly there is a higher

TABLE 3.9

Per caput consumption of starchy food in eight Southeast Asian countries (1964-66 average)[1]

(thousand tonnes otherwise specified)

Country and food	Production	Importation	Available supply	Food	Per caput consumption				
					kg/yr	g/day	Cal/day Number	Protein per day/g	Fat per day/g
Cambodia									
Sweet potato	21		21	17	2.7	7.4	7	0.1	-
Cassava	21		21	19	3.0	8.3	9	0.1	-
Total, root crops					5.7	15.7	16	0.2	-
Banana	140		140	126	20.1	55.1	37	0.5	0.2
Grand total					25.8	70.8	2231	57.7	19.9
Indonesia									
Sweet potato	2987		2987	2689	25.4	69.7	68	0.8	0.2
Cassava	11114		11114	9080	85.9	235.3	256	2.1	0.5
Potato	42		42	32	0.3	0.8	1	-	-
Others					1.3	3.7	13	-	-
Total, root crops					112.9	309.5	338	2.9	0.7
Banana					0				
Grand total							1760	38.4	22.8
Laos									
Sweet potato	14		14	12	4.4	12.1	12	0.1	-
Cassava	10		10	8	3.1	8.5	9	0.1	-
Total, root crops					7.5	20.6	21	0.2	-
Banana	2	1	3	3	1.0	2.8	2	-	-
Grand total					8.5	23.4	2036	44.6	13.6
Malaysia									
Sweet potato	100		100	30	3.7	10.2	10	0.1	-
Cassava/flour	12		12	12	1.5	4.1	14	0.1	-
Potato		21	21	20	2.5	6.8	5	0.1	-
Others					0.7	2.0	7	-	-
Total, root crops					8.4	23.1	36	0.3	
Banana	330		330	243	30.2	82.8	55	0.7	0.2
Grand total					38.2	105.9	2200	49.4	40.8
Philippines									
Sweet potato	717		717	659	20.4	55.8	54	0.6	0.2
Cassava	596		596	443	13.7	37.6	41	0.3	0.1

(cont.)

TABLE 3.9 *(cont.)*

Per caput consumption of starchy food in eight Southeast Asian countries (1964-66 average)[1]

(thousand tonnes unless otherwise specified)

Country and food	Production	Importation	Available supply	Food	Per caput consumption kg/yr	g/day	Cal/day Number	Protein per day/g	Fat per day/g
Potatoes	17		17	13	0.4	1.1	1	-	-
Others					4.8	13.2	7	0.1	-
Total					39.3	107.7	112	1.0	0.3
Banana	984		984	935	28.9	79.2	53	0.7	0.2
Grand total					68.2	186.9	1911	43.8	31.4
Singapore									
Sweet potato	5		5	5	2.7	7.3	7	0.1	-
Cassava/flour	3		3	3	1.6	4.4	15	0.1	-
Potatoes		31	31	17	9.1	25.0	17	0.4	-
Others					6.4	17.7	59	0.1	-
Total					19.8	54.4	98	0.7	-
Banana		22	22	20	10.7	29.4	20	0.3	0.1
Grand total					30.5	83.8	2443	63.3	48.9
Thailand									
Sweet potato	202		202	182	5.9	16.2	15	0.3	-
Cassava/flour	331		331	170	5.5	15.1	51	0.2	0.1
Potatoes	6		6	4	0.1	0.4	-	-	-
Total					11.5	31.7	66	0.5	0.1
Banana	1063			952	31.0	84.9	57	0.8	0.3
Grand total			1063		42.5	116.6	2226	50.9	27.3
Viet Nam									
Sweet potato	275		275	194	12.0	32.9	32	0.6	-
Cassava	268		268	214	13.3	36.4	40	0.3	-
Potatoes	7		6	5	0.3	0.8	1	-	-
Total					25.6	70.1	73	0.9	-
Banana	207		206	185	11.5	31.5	21	0.3	0.1
Grand total					37.1	101.6	2200	48.6	21.4

[1]FAO, Food Balance Sheets, 1971.
Source: Oñate *et al.*, 1976.

TABLE 3.10

Average daily per caput nutrient intake in percentage of recommended allowance[1] in eight regions in the Philippines[2]

	Metropolitan Manila	Ilocos Mountain Province	Batanes Cagayan Valley	Southern Tagalog	Eastern Visayas	Western Visayas	Northeastern Mindanao	Southwestern Mindanao	Weighted mean
Households	(402)	(274)	(293)	(368)	(306)	(512)	(187)	(225)	
Nutrient									
Calories	82	88	81	79	68	75	81	74	76
Protein	92	97	87	81	80	88	89	86	86
Calcium	36	37	39	31	30	42	30	31	34
Iron	130	150	150	100	112	125	125	129	125
Vitamin A	60	79	64	39	32	59	43	51	50
Thiamine	73	87	65	64	56	74	67	44	64
Riboflavin	53	49	41	37	27	36	33	29	36
Niacin	125	147	115	117	92	125	136	118	117
Ascorbic acid	89	121	117	83	114	90	77	75	90

[1]Values were rounded off for ease in presentation.
[2]Source of data: Food and Nutrition Research Center Surveys.
Source: Oñate *et al.*, 1976.

TABLE 3.11

Average daily per caput food intake in percentage of recommended allowance[1] in eight regions in the Philippines[2]

Households	Metropolitan Manila (402)	Illocos Mountain Province (274)	Batanes Cagayan Valley (293)	Southern Tagalog (368)	Eastern Visayas (306)	Western Visayas (512)	Northeastern Mindanao (187)	Southwestern Mindanao (225)	Weighted mean
1. Cereals	107	122	107	107	95	107	111	110	107
2. Starchy roots and tubers	8	73	77	76	202	45	64	45	73
3. Sugars and syrups	70	46	62	84	34	47	70	45	54
4. Dried beans, nuts and seeds	20	118	76	41	18	53	29	31	37
5. Leafy and yellow vegetables	22	36	35	13	25	29	31	33	26
6. Vitamin C-rich foods	51	47	42	31	9	27	10	20	26
7. Other fruits and vegetables	137	115	170	116	42	99	82	72	93
8. Meat, poultry and fish	94	71	68	63	67	66	75	71	70
9. Eggs	92	40	33	40	14	20	20	20	27
10. Milk and milk products	70	12	12	15	3	12	8	10	14
11. Fats and oils	39	21	28	37	14	18	30	18	24

[1]Values were rounded off for ease in presentation.
[2]Source of data. Food and nutrition Research Center Surveys.
Source: Oñate *et al.*, 1976

consumption in the urban areas than in the rural areas, indicating the importance of yam as an item of food and reflecting the inflated market price of this preferred root crop as a result of limited production (See Table 3.12).

In Nigeria, although cassava as dry *gari* is consumed more in the urban areas than in rural areas the reverse is true with yam, probably because of the expense of transporting fresh yams and the ease of preparing meals using dried *gari*, which is of great convenience to urban workers (see Table 3.13).

Zones where root crops are consumed do not necessarily coincide with a high incidence of malnutrition. The Indian state of Kerala may serve as an example. This state has a population of about 25 million, whose staple food is rice. However, because of the high population density, fertile land suitable for the cultivation of rice is now in short supply and so most of the rice is cultivated on the poorly drained but fertile lowlands, while the well-drained hilly areas of low fertility are planted mainly with cassava. The main staples are therefore rice and cassava.

As the population increased rapidly there was less land for the cultivation of rice and the production, yield and consumption of cassava increased.

This might have been expected to adversely affect nutritional status. Taking infant mortality as an index of nutritional status provides reassuring evidence, as the infant mortality in Kerala remained relatively low. Table 3.14 shows that in 1970/71 cassava supplied over 740 calories out of a total daily calorie intake of 2 519, which is probably adequate. Protein intake was less than 40 g per day. Cassava supplied very little but some of this deficit was made up by ingestion of rice and fish. So by varying the diet to include some cereal and animal protein, root crops like cassava are very useful in supplementing the energy obtained from cereals.

This has been confirmed by balance sheet data discussed by Goering (1979) indicating that serious protein deficiency is not necessarily common in countries where root crops are one of the sources of calories. Thus in ten African countries where root crops supply between 500 to 900 calories or 20 to 40 percent of the total daily calorie intake, seven exhibited a per caput calorie consumption level of under 2 200 cal/day and only one showed an

TABLE 3.12

Food consumption in Ghana in 1961-62 (g/caput/day)

Food		Urban	Rural	Urban as percentage of rural
Maize	dry grain	10.7	61.1	17.5
	dough	41.2	67.8	60.8
Millet		59.0	44.2	133.5
Guinea corn		14.3	11.6	123.3
Koko[1]		28.7	6.0	478.3
Rice		21.9	36.6	59.8
Bread		15.9	8.6	184.9
Cassava	fresh roots	112.2	196.1	57.2
	gari	15.6	16.2	96.3
Plantain		193.5	119.4	162.1
Cocoyam		72.3	44.7	161.7
Yam		110.6	51.6	214.3
Fish	fresh	6.0	11.6	51.7
	smoked	22.0	20.1	109.4
Meat	fresh	41.0	15.2	269.7
	preserved	2.0	2.0	100
Fats and oils		11.7	4.3	272.1
Sugar		4.0	2.1	190.4

[1]*Koko* is a starchy paste or pap prepared from cereal or root flour.
Source: Calculated from Whitby, P., *A review of information concerning food consumption in Ghana.* FAO, Rome. 1969.

intake greater than 2 400 calories, yet none had a protein intake of less than 40 g per day, and only three had less than 50 g per day. Thus a limited intake of calories from root crops is not necessarily inconsistent with adequate protein intake.

Besides root crops, the leaves of cassava, sweet potato and cocoyam are commonly consumed in many tropical countries including Zaire, Papua New Guinea and central Java in Indonesia, especially during periods of food shortage. These leaves contribute some protein to the diet. They also contain minerals, particularly iron and calcium, and provide a valuable

TABLE 3.13
Food consumption in Nigeria (g/caput/day)

Food	Rural[1]	Urban[2]	Urban as a percentage of rural
Yam, fresh tuber	287.8	70.0	24
Cassava, dry *gari*	43.1	141.0	327
Cocoyam, fresh	33.8	-	
Irish potato	31.8	-	
Plantain, boiled fruit	13.5	9.0	68
dry flour	10.3	-	
Taro, boiled	16.7	-	
Maize, meal	162.8	-	
grain	27.3	-	
dry starch	17.0	36.0	211
Millet, meal	88.8	4.0	
fura	16.8	-	
Guinea corn, meal	16.4	-	
Acha, grain	22.0	-	
Rice	11.7	47.0	401
Wheat	1.3	31.0	2 384
Cowpea	21.9	33.0	150
Locust bean	13.8	-	
Beef	23.3	35.0	150
Fish, dry	3.5	5.0	142
Red palm oil	27.7	20.0	72
Cows milk, fresh	35.2	6.0	17
Sugar	4.4	5.0	113
Fish, fresh	-	58.0	5 800
Egg	-	4.0	400

[1]Rural figures based on studies by Collis, Dema, Lesi & Omololu (1962).
[2]Urban figures calculated on the basis of a study by McFie (1967) in Lagos.

TABLE 3.14
Food consumption in Kerala, 1970/71
(average daily per caput value)

Food	Total consumption (g)	Calories	Protein (g)
Rice	289	1 000	18.5
Cassava (tapioca)	474	744	3.3
Coconuts	60	267	2.7
Fruit	87	68	0.7
Fish	41	46	8.3
Milk	30	23	1.0
Meat	5	6	1.1
Oil	24	212	-
Sugar	25	100	-
Subtotal		2 466	35.6
All other		53	2.2
Total		2 519	37.8

Source: United Nations, 1975.

source of vitamins A and C. Increased consumption of these green leaves could help reduce the incidence of xerophthalmia in countries where nutritional blindness is prevalent. The use of edible cassava leaves as a green vegetable is popular in Africa (Hahn, 1984).

4. Nutritive value

The main nutritional value of roots and tubers lies in their potential ability to provide one of the cheapest sources of dietary energy, in the form of carbohydrates, in developing countries. This energy is about one-third of that of an equivalent weight of grain, such as rice or wheat, because tubers have a high water content. However, the high yields of most root crops ensure an energy output per hectare per day which is considerably higher than that of grains (see Table 4.1). Sweet potato for example has a tremendous capacity for producing high yields, up to 85 t/ha have been recorded on experimental plots, though most plantation yields do not exceed 20 t/ha. As shown in Table 4.1, potato is one of the highest calorie-yielding crops in the world. Such root crops are particularly valuable in the tropics where most of the population depends on carbohydrate foods as dietary staples.

Because of the low energy content of root crops compared to cereals on a wet basis, it is often assumed that root crops are not suitable for use in baby foods. This is not necessarily true if their energy density is increased by drying. Tapioca, for instance, is used in a number of commercial baby foods in industrialized countries. Composite flours prepared from root crops and cereals could be used in baby food formulas, if appropriately supplemented. The addition of germinated (malted) cereals to cassava flour increases the energy density of gruels prepared from it, by reducing their viscosity through the action of amylolytic enzymes. However, the use of fresh cassava products as infant weaning foods should be discouraged, because of probable toxicity, low protein content and energy density. Infants and young children, pregnant and lactating women are among the most nutritionally vulnerable people. Their nutrient requirements are specifically higher in order to meet the increased physiological demand for growth and lactation. These requirements are listed in Tables 4.2 and 4.3 together with those for adolescents and adults.

TABLE 4.1

Comparison of average energy and protein production of selected food crops in developing countries (per hectare and per day)

Crop	Growth duration (days)	Dry matter (kg/ha/day)	Edible energy ('000 kcal/ ha/day	Edible protein (kg/ha/day)	Production value (US$/ ha/day
Potato	130	18	54	1.5	12.60
Yam	180	14	47	1.0	8.80
Sweet potato	180	22	70	1.0	6.70
Rice, paddy	145	18	49	0.9	3.40
Groundnut in shell	115	8	36	1.7	2.60
Wheat	115	14	40	1.6	2.30
Lentil	105	6	23	1.6	2.30
Cassava	272	13	27	0.1	2.20

Source: FAO, *Production yearbook 1983* (Rome, 1984); USDA, *Composition of foods* (Washington, D.C. 1975); and FAO, *Report of the agroecological zones project* (Rome, 1978). Production estimates are 1981-83 averages; price estimates are for 1977.
Adapted and modified from Horton *et al.*, (1984).

Undernutrition is often the outcome of either an insufficient food intake or poor utilization of food by the body, or both simultaneously. Recent surveys show that very few people in tropical countries suffer from a simple protein deficiency. The most prevalent deficiency is protein-energy, in which an overall energy deficiency forces the metabolism to utilize the limited intake of protein as a source of energy. This is an area in which root crops could play a more significant role as additional sources of dietary energy and protein. Increasing the consumption of root crops could help save the much-needed protein provided essentially by other foods such as cereals and legumes. Traditionally, in Africa, root crops such as cassava are eaten with a soup or stew made of fish, meat or vegetables, providing an excellent supplement to cassava meal.

TABLE 4.2

Average daily energy, protein, vitamin A, folic acid, iron and iodine requirements for infants and children

Age	Median weight (kg)	Energy[1] (kcal)		Protein[1] (g)	Vitamin[2] A (μg)	Folic[2] acid (μg)	Iron[2] (mg)	Iodine[3] (mg)
Infants (months)								
3-6	7.0	700		13.0	350	25	14	40
6-9	8.5	810		14.0	350	31	14	50
9-12	9.5	950		14.0	350	34	14	50
Children (years)								
1-2	11.0	1 150		13.5	400	36	8	70
1-3	13.5	1 350		15.5	400	46	9	70
3-5	16.5	1 550		17.5	400	54	9	90
		boys	girls					
5-7	20.5	1 850	1 750	21.0	400	68	9	90
7-10	27.0	2 100	1 800	27.0	400	89	16	120

Notes:

[1]Values derived from *Energy and protein requirements: report of a joint FAO/WHO/UNU expert consultation.* WHO Technical Report Series 724. Geneva, 1985.

[2]Values derived from *Requirements of vitamin A, iron, folate and vitamin B_{12}: report of a joint FAO/WHO Expert Consultation.* (In press)

[3]Values derived from *Recommended dietary allowances:* Ninth rev. ed., US National Academy of Sciences. Washington, D.C., 1980.

Source: FAO, 1988b.

TABLE 4.3

Average daily energy, protein, vitamin A, folic acid, iron and iodine requirements for adolescents and adults

Age (years)	Median weight (kg)	Energy[1] (kcal)	Protein[1] A (g)	Vitamin[2] acid (µg)	Folic[2] (µg)	Iron[2] (mg)	Iodine[3] (mg)
Males							
10-12	34.5	2.200	34.0	500	102	16	150
12-14	44.0	2 400	43.0	600	170	24	150
14-16	55.5	2 650	52.0	600	170	24	150
16-18	64.0	2 850	56.0	600	200	15	150
>18	70.0	3 050	52.5	600	200	15	150
Females							
10-12	36.0	1 950	36.0	500	102	16	150
12-14	46.5	2 100	44.0	600	170	27	150
14-16	52.0	2 150	46.0	600	170	27	150
16-18	54.0	2 150	42.0	500	170	29	150
>18	55.0	2 350	41.0	500	170	29	150
Pregnant							
full activity		+285	+6.0	600	370-470	47[4]	+25
reduced activity		+200	+6.0	600	370-470	47[4]	+25
Lactating							
first 6 months		+500	+17.5	850	270	17	+50
after 6 months		+500	+13.0	850	270	17	+50

Notes:

[1]Values derived from *Energy and protein requirements: report of a joint FAO/WHO/UNU expert consultation.* WHO Technical Report Series 724. Geneva, 1985.
[2]Values derived from *Requirements of vitamin A, iron, folate and vitamin B_{12}: report of a joint FAO/WHO Expert Consultation.* (In press)
[3]Values derived from *Recommended dietary allowances.* Ninth rev. ed., US National Academy of Sciences. Washington, D.C., 1980.
[4]Among pregnant women, dietary supplementation of iron is usually called for because the iron requirement cannot be met through normal dietary intake.
+In addition to the normal requirement.
Source: FAO, 1988b.

NUTRITIONAL COMPOSITION OF ROOTS AND TUBERS

As with all crops, the nutritional composition of roots and tubers varies from place to place depending on the climate, the soil, the crop variety and other factors. Table 4.4 shows the nutritional composition for common roots and tubers and the amino-acid composition of some root crop proteins along with a comparison of suggested amino-acid requirement is shown in Table 4.5.

The main nutrient supplied by roots and tubers is dietary energy provided by carbohydrates. The protein content is low (one to two percent) and in almost all root crop proteins, as in legume proteins, sulphur-containing amino-acids are the limiting amino-acids (Tables 4.5, 4.9). Cassava, sweet potato, potato and yam contain some vitamin C and yellow varieties of sweet potato, yam and cassava contain beta-carotene or provitamin A. Taro is a good source of potassium. Roots and tubers are deficient in most other vitamins and minerals but contain significant amounts of dietary fibre. Leaves of taro are cooked and eaten as a vegetable. They contain beta-carotene, iron and folic acid, which protects against anaemia. Leaves of sweet potato and cassava are also commonly eaten.

CARBOHYDRATES

The dry matter of root crops, banana and plantain is made up mainly of carbohydrate, usually 60 to 90 percent. Plant carbohydrates include celluloses, gums and starches, but starches are the main source of nutritive energy as celluloses are not digested.

Starches are made up of two main polymers, a straight chain glucose polymer called amylose, which usually constitutes about 10 to 30 percent of the total, and the branched chain glucose polymer, amylopectin, which makes up the rest. The principal constituent of edible carbohydrate is starch together with some sugars, the proportion depending on the root crop.

The physical properties of starch grains influence the digestibility and processing qualities of root crops. The starch granules of some varieties of cocoyam are very small, about one-tenth those of potato, which improves the starch digestibility, making these varieties more suitable for the diets of infants and invalids. For the preparation of certain foods like *fufu*, a stiff dough is required and so the rheological properties of the starch paste become significant. The viscosity of starch-water pastes of different yam starches varies considerably from a relatively low value for *D. dumetorum* through increasing viscosity in *D. esculenta* to the highest value in *D. rotundata* (see Table 4.6). Hence *D. rotundata* is traditionally the accepted yam for *fufu*. Most yams give viscous pastes with a much higher gel strength than that of other crops. Therefore yams are traditionally preferred for *fufu*, a starch paste which is prepared by pounding cooked roots or tubers in a mortar with a pestle (Rasper, 1969, 1971). Cassava starch has some special characteristics for food processors. It is readily gelatinized by cooking with water and the solution after cooling remains comparatively fluid. The solutions are relatively stable and do not separate again into an insoluble form (retrogradation) as is the case with maize and potato starch.

In addition to starch and sugar, root crops also contain some non-starch polysaccharides, including celluloses, pectins and hemicelluloses, as well as other associated structural proteins and lignins, which are collectively referred to as dietary fibre (Table 4.7). The role of dietary fibre in nutrition has aroused a lot of interest in recent years. Some epidemiological evidence suggests that increased fibre consumption may contribute to a reduction in the incidence of certain diseases, including diabetes, coronary heart disease, colon cancer, and various digestive disorders. The fibre appears to act as a molecular sieve, trapping carcinogens which would otherwise have been recirculated into the body; it also absorbs water thus producing soft and bulky stools. Sweet potato is a significant source of dietary fibre as its pectin content can be as high as 5 percent of the fresh weight or 20 percent of the dry matter at harvest (Collins and Walter, 1982). However, banana, which is also known to have a beneficial effect in correcting intestinal disorders, appears to contain very little dietary fibre, only 0.84 percent using traditional methods of analysis. Because of this, Forsythe (1980) carried out some studies on the cell wall materials of banana pulp by extracting with ascorbic acid, centrifuging and washing the sugars away. The residue, comprising 3.3 percent of the pulp had a water-holding capacity 17 times its dry weight. Analysis yielded 15.2 percent lignin, 13 percent starch, 9.8 percent protein, 4.8 percent cellulose, 3.7 percent lipid, 1.3

TABLE 4.4

Nutritive values of tropical root crops (per 100 g edible portion)

Crop	Food energy Kj	Moisture (%)	Protein (g)	Fat (g)	Protein/ energy ratio (g/1 000 Kcal)	Fibre (g)	Total CHO & fibre (g)	Ash (g)	Ca (mg)	P (mg)	Fe (mg)	K (mg)	Na (mg)	Carotene equi. (µg)	Thiamin (mg)	Riboflavin (mg)	Niacin (mg)	Ascorbic acid (mg)	Folic acid (µ)
Cassava	565	65.5	1.0	0.2	7	1.0	32.4	0.9	26	32	0.9	394	2	0	0.05	0.04	0.6	34	24.2
Sweet potatoes																			
white	452	72.3	1.0	0.3	15	0.8	25.1	0.7	21	50	0.9	210	31	35	0.14	0.05	0.7	21	52.0
yellow	481	70.0	1.2	0.3	-	0.8	27.1	0.7	36	56	0.9	304	36	1 680	0.12	0.05	0.6	30	-
Potato	335	78.3	2.0	0.1	27	0.4	19.1	0.9	9	55	0.7	451	7	30	0.11	0.04	1.2	14	-
Yam	452	71.8	2.0	0.1	21	0.5	25.1	1.0	22	39	1.0	294	10	0	0.10	0.04	0.07	-	-
Taro and tamia	393	75.4	2.2	0.4	-	0.8	21.0	1.0	34	62	1.2	448	10	tr.	0.12	0.04	1.0	8	-
Giant taro	255	83.0	0.6	-	-	-	14.8	-	30	50	1.0	-	-	0	0.05	-	-	5	-
Giant swamp taro	548	-	0.9	-	-	-	31.0	-	334	56	1.2	-	-	0	0.05	0.07	0.88	-	-
Elephant yam	339	78.5	2.0	-	-	-	18.4	-	38	38	2.4	416	-	0	0.06	0.02	1.7	6	-
Taro leaves	255	81.4	4.0	-	-	-	11.9	-	162	69	1.0	963	-	5 535	0.13	0.34	1.5	63	163.0
Sweet potato tips	-	86.1	2.7	-	-	-	-	-	74	-	4	-	-	5 580	0.32	-	-	41	-
Banana	425	71.6	1.2	0.3	-	0.6	26.1	0.8	12	32	0.8	401	4	225	0.03	0.04	0.6	14	-
Plantain	476	68.2	0.9	0.2	-	0.4	29.7	1.0	19	38	0.6	-	-	475	0.15	0.06	0.7	11	-

Source: FAO, 1972.

TABLE 4.5

Comparison of suggested amino-acid requirement patterns with amino-acid composition of tropical root crops

Amino acid (mg/g crude protein)	Suggested pattern of requirement					Reported composition								
	Infant mean	FAO ref.	Preschool child (2-5 yr)	School child (10-12 yr)	Adult	Egg	Cow's milk	Beef	Potato	Sweet potato	Cassava	Cocoyam	Banana	Yam
His	26	-	19	19	16	22	27	34	20	13	21	18	75	19
He	46	42	28	28	13	54	47	48	39	37	28	35	29	37
Leu	93	49	66	44	19	86	95	81	59	54	40	74	47	65
Lys	66	42	58	44	16	70	78	89	60	34	41	39	41	41
Met + Cys	42	40	25	22	17	57	33	40	30	28	27	40	47	28
Phe + Tyr	72	56	63	22	19	93	102	80	78	62	41	87	65	80
Thre	43	28	34	28	9	47	44	46	39	38	26	41	34	36
Tryp	17	14	11	9	5	17	14	12	14	14	12	14	13	13
Val	55	42	35	25	13	66	64	50	51	45	33	61	40	47
Total														
Incl. his	460	-	339	241	127	512	504	479	382	325	269	409	391	366
Excl. his	434	313	320	222	111	490	477	445	363	312	248	391	316	347
Chemical score	-	-	-	-	-	100	100	80	53	82	85	70	71	75
Limiting amino-acids								Val	SAA	SAA	AAA	AAA	AAA	SAA

Notes: Chemical score = $\frac{\text{mg amino-acid in test protein}}{\text{mg amino-acid in requirement pattern}} \times 100$

Chemical scores of potato: Infant = 63; Preschool child = 90; School child/Adult = >100.
SAA = Sulphur amino-acids (Met + Cys)
AAA = Aromatic amino-acids (Phe + Tyr).

Source: Data from World Health Organization (1985)

TABLE 4.6
Rheological properties of various yam starches

Species and cultivar	Pasting temp. °C	Viscosity (Brabender units)		Gel strength (ml) after		
		(on attaining 95°C)	(maximum reached before cooling)	24 h	96 h	168 h
D. rotundata						
Puna	76	450	630	8.8	13.6	14.1
Labreko	78-79	260	470	4.3	6.2	8.0
Kplinjo	77	330	490	10.6	12.7	13.3
Tantanpruka	79	610	650	12.4	17.2	20.5
Tempi	80-82	430	520	7.5	10.5	10.8
D. alata						
White fleshed	85	25	110	14.8	16.5	17.2
Purple fleshed	81	80	200	14.8	18.5	19.4
D. esculenta	82	25	55	2.5	4.0	4.6
D. dumetorum	82	25	25	-	-	-

Source: Rasper and Coursey (1967).

TABLE 4.7
Fibre as percentage of dry matter in raw sweet potato and banana

	Sweet potato	Banana
Cellulose	3.26	1.0
Hemicellulose	4.95	5.8
Insoluble pectin	0.50	-
Lignin	NR	0.2

TABLE 4.8

Calorie and protein contribution of starchy staples to diets in developing country regions,
(in percentage of regional total) **1979-81**[1]

	Temperate S. America	Tropical S. America	Central America	East/South Africa	Equatorial Africa	Humid W. Africa	Semi-Arid W. Africa	N. Africa M. East	India	S. Asia	S.E. Asia	China	All regions
Calories													
Cassava	0.2	5.7	0.5	18.1	31.4	16.0	8.4	-	0.9	0.5	4.0	0.2	2.6
Potato	3.3	2.0	0.7	0.7	0.7	0.1	0.1	1.3	0.9	0.8	0.5	0.7	0.9
Sweet potato	0.6	0.3	0.5	1.6	4.5	0.5	0.3	0.1	0.3	0.9	1.5	11.1	3.9
Yam	-	0.1	0.1	-	1.1	9.5	9.4	0.1	-	-	-	-	0.5
Aroids	-	-	-	-	0.3	3.4	1.3	-	-	-	-	0.1	0.1
Others	-	0.1	0.1	-	0.8	-	0.1	0.1	-	0.1	0.2	-	0.1
Plantain/banana	0.6	3.7	2.1	2.6	7.6	5.7	1.3	0.1	0.4	1.4	1.4	-	1.0
All above	4.7	11.9	4.0	23.0	46.4	35.2	20.9	1.7	2.5	3.7	7.6	12.1	9.1
Protein													
Cassava	-	1.3	-	5.9	8.2	5.3	1.6	-	-	-	1.4	-	-
Potato	2.4	1.0	-	-	-	-	-	0.9	-	-	-	-	1.0
Sweet potato	-	-	-	-	2.6	-	-	-	-	-	-	5.0	1.7
Yam	-	-	-	-	-	6.6	6.1	-	-	-	-	-	-
Aroids	-	-	-	-	-	2.2	2.0	-	-	-	-	-	-
Others	-	-	-	-	-	-	-	-	-	-	-	-	-
Plantain/banana	-	1.3	-	-	2.1	1.8	-	-	-	-	-	-	-
All above	2.4	3.6	-	5.9	12.9	15.9	9.7	0.9	-	-	1.4	5.0	2.7

[1]Corrected by the amino-acid protein quality.
Source: FAO, 1987d.

percent pectin and 0.4 percent ash. There is therefore a need to pay more attention to the significance of fibrous components in these root crops, especially in banana and sweet potato, and to determine their composition and dietary function. Other root crops, particularly yam, contain mucilages, which have a considerable influence on their cooking qualities.

PROTEIN

The protein content and quality of roots, tubers, bananas and plantains are variable; that of yam and potato is highest, being approximately 2.1 percent on a fresh weight basis. The protein contribution of these foods to the diet in developing countries, corrected by the amino-acid protein quality is, on a worldwide average, only 2.7 percent, provided mainly by potato and sweet potato. However these starchy staples do provide a much greater proportion of the protein intake in Africa (Table 4.8), ranging from 5.9 percent in East and southern Africa to a maximum of 15.9 percent in humid West Africa, supplied mainly by yam and cassava. These figures do not include the protein contribution from the leaves of crops such as cassava, sweet potato and cocoyam which are eaten as green vegetables. The amino-acid content of roots and tubers, unlike most cereals, is not complemented by that of legumes as both are limiting in respect of the sulphur amino-acids (see Table 4.9). In order to maximize their protein contribution to the diet, roots and tubers should be supplemented with a wide variety of other foods, including cereals.

To some extent the protein content of root crops is influenced by variety, cultivation practice, climate, growing season and location (Woolfe, 1987). In potato, the addition of nitrogen fertilizer increases the protein content (Eppendorfer *et al.*, 1979; Hoff *et al.*, 1971) while in the case of sweet potato the protein content could vary from 2.0 to 7.5 percent depending on the cultivar and treatment. Nitrogen fertilizer increases the protein content of sweet potato, but the lysine content is decreased, while the aspartic acid and free amino-acids are increased (Yang, 1982). Also leafy growth is increased at the expense of tuber production.

In root crops the quality of the protein, in terms of the balance of essential amino-acids present, may be compared to that of standard animal proteins in beef, egg or milk (see Table 4.5). Most root crops contain a reasonable amount of lysine, though less than in legumes, but the sulphur amino-acids are limiting. For example, yam is rich in phenylalanine and threonine but limiting in the sulphur amino-acids, cystine and methionine and in tryptophan.

Protein quality may be assessed in terms of the amino-acid score but the biological utilization of protein depends also on the composition of the diet, the protein digestibility and the presence of toxins or other antinutritional factors. This is reflected in the net protein utilization (NPU) proportions of nitrogen intake that is retained or biological value (BV) of the protein, which estimates the proportion of absorbed nitrogen that is retained (Table 4.10) either by measurement of nitrogen balance, or preferably by direct studies on experimental animals. Results may also be expressed as protein efficiency ratios (PER values) where PER = gain in weight in grams divided by the protein intake in grams.

In feeding studies conducted on rats, banana proteins were utilized as well as those of maize, although their utilization was less efficient than those of yam, cocoyam and sweet potato. The protein of potato is of good nutritional quality with a relatively high lysine content, and so it can be used in developing countries to complement foods low in lysine. As shown in Table 4.10, its utilizable protein as a percentage of its calorie content is as high as that of wheat.

The protein of sweet potato is also of acceptable nutritive value, with a chemical score of 82 and sulphur amino-acids as the major limiting factors. The quality of the protein will depend on the severity of heat treatment during the processing of sweet potato products. (Walter *et al.*, 1983). Horigone *et al.*, (1972) reported a PER of 1.9 for a protein isolated from a sweet potato starch production factory. This value could be increased to 2.5 by the addition of lysine and methionine, indicating a deficiency of methionine and the destruction of lysine during processing. When unheated sweet potato flour was added to wheat in the diet of rats at the 30 percent level, the biological value of the diet was increased from 72 to 80 owing to the improved protein value. A similar result was obtained when sweet potato flour replaced rice (Yang, 1982). Walter and Catignani

(1981) extracted a white protein isolate and a greyish-white protein concentrate (chromoplast protein) from two sweet potato varieties, "Jewel" and "Centennial" and found that they gave a very good amino-acid pattern, with lysine higher than the FAO pattern (Table 4.11). Both the isolates gave a higher gain in weight and a better PER than casein, though this was not statistically significant, indicating that some protein fractions from selected varieties of sweet potato are of very high quality (Yang, 1982).

Cassava protein is lower in total essential amino-acids than the other root crops but recently Adewusi *et al.* (1988) found that cassava flour used as a component in animal feeding trials was a more effective replacement for wheat than either sorghum or maize. The content of protein in yam varies between 1.3 and 3.3 percent, (Francis *et al.,* 1975), but based on the quantity consumed by an adult in West Africa, about 0.5 to 1 kg per caput/day, it can contribute about six percent of the daily protein intake (see Table 4.8). The chemical score for yam proteins, using the FAO reference protein as standard, varied from 57 to 69 (Francis *et al.,* 1975). The incidence of kwashiorkor has been reported to be high in yam-consuming areas. This emphasizes the need to supplement a yam-based diet with more protein-rich foods in order to support active growth in infants. Fresh cocoyam contains a high percentage of water and is a food of low-energy density compared to alternative root crops. It has a protein content of about two percent (Table 4.4) with a chemical score of 70 (Table 4.5). However chemical score alone is not a satisfactory index of protein availability and efficiency in the diet. This can best be assessed by controlled feeding trials to obtain values of digestibility. Such values have been determined for many individual foods. If information is not available on the digestibility of the protein in a particular diet, the value can be estimated by using values for individual components and calculating a weighted mean according to the proportion of protein supplied by these foods. In foods of low protein content such as yam and cassava, feeding trials to determine the biological efficiency of the protein are often inconclusive. As an approximate correction, for a diet based on vegetable protein, a digestibility factor of 85 percent may be applied (WHO, 1985).

TABLE 4.9

Essential amino-acids of plantain, cassava, sweet potato, cocoyam and yam compared with cowpea

Amino-acids (mg N/g)	Plantain	Cassava	Sweet potato	Cocoyam	Yam	Cowpea
Lysine	193	259	214	241	256	427
Threonine	141	165	236	257	225	225
Tyrosine	89	100	146	226	210	163
Phenylalanine	134	156	241	316	300	323
Valine	167	209	283	382	291	283
Tryptophan	89	72	-	88	80	68
Isoleucine	116	175	230	219	234	239
Methionine	48	83	106	84	100	73
Cystine	65	90	69	163	72	68
Total sulphur-containing	113	173	175	247	172	141
Total	1 042	1 309	-	1 976	1 768	1 869

Source: FAO, 1970.

TABLE 4.10

Utilizable protein in some staple foods (percentage of calories)

	Total protein	Utilizable protein
Sago	0.6	0.3
Cassava	1.8	0.9
Plantain	3.1	1.6
Yam	7.7	4.6
Maize	11.0	4.7
Rice	9.0	4.9
Potato	10.0	5.9
Wheat	13.4	5.9

Source: Payne, 1969.

TABLE 4.11

Comparison of essential amino-acid patterns for chromoplast and white protein in Jewel and Centennial sweet potato roots to the FAO reference protein

Amino-acid[1]	Chromoplast		FAO	White	
	Jewel	Centennial		Jewel	Centennial
Threonine	5.77	5.67	4.0	6.43	6.39
Valine	7.83	7.68	5.0	7.90	7.89
Methionine	2.26	2.10		2.03	1.84
Isoleucine	6.01	5.89	4.0	5.63	5.71
Leucine	9.64	8.95	7.0	7.40	7.44
Tyrosine	6.71	6.41	6.0	6.91	7.09
Phenylalanine	7.08	7.15		8.19	7.94
Lysine	7.03	6.43	5.5	5.16	5.21
Tryptophan	1.56	1.77	1.0	1.23	1.44
PER	2.73	2.78		2.64	2.63

[1]g amino-acid/16 g N
Source: Walter and Catignani, 1981.

Human dietary tests have been carried out using root crops to test the efficiency of the root crop protein to maintain good health in the absence of other protein sources. Most of this work has been done on potato and is well documented by Woolfe (1987). The classical work of Rose and Cooper (1907) indicated that young women could be maintained in nitrogen balance for seven days on a diet in which potato supplied 0.096 g N/kg body weight. This has been confirmed more recently in experiments in which a potato protein level of 0.0545 g/kg body weight was found to maintain nitrogen balance in healthy college students, compared to a value of 0.0505 g/kg body weight obtained for egg.

Lopez de Romana *et al.* (1981) in Peru reported that potato can be used successfully to supply up to 80 percent of the daily requirement of protein and 50 to 75 percent of the energy of infants and young children if the remaining energy and nitrogen is provided by a non-bulky, easily digestible food. Acceptability, digestibility, tolerance and growth of children were analysed. Excellent acceptability and tolerance were found for a diet providing about 50 percent of the energy from potato with casein added to make up to 80 percent of the total dietary energy from protein. Raising the level of potato to provide 75 percent of the dietary energy tends towards poor acceptability and tolerance near the last week of the three-month study mainly because of the bulk and the poor digestibility of the carbohydrates.

When the British settled on the remote South Pacific Island of Tristan da Cunha in 1876, it was reported in 1909 that the population had increased and were very healthy on a potato-based diet, consuming about 3-4 lb of potatoes per day (Kahn 1985). Even in an affluent country such as the United Kingdom, potato contributed about 3.4 percent of the total household protein intake according to the National Food Survey Committee (1983), compared to 1.3 percent for fruit, 4.6 percent for egg, 4.8 percent for fish, 5.8 percent for cheese, 5.7 percent for beef, 9.8 percent for white bread and 14.6 percent for milk.

In dietary tests adult Yami tribesmen were given a diet based on sweet potato supplemented with fish and vegetables, designed to supply 0.63 g protein/kg body weight/day. They did not show any physical abnormality after two months, but appeared to tire more easily after a more prolonged period on this diet. As a result of the high dietary fibre content the faecal volume of the test subjects was very high, an average of 800 g on a wet weight basis per day. This diet, contrary to expectation, did not generally reduce the serum cholesterol and total lipids, as did some other vegetables, though a particular sweet potato variety did significantly reduce these factors (Yang, 1982).

However, when seven teenage boys were placed on two similar diets based on sweet potato, supplying 0.67 g protein and 0.71 g protein/kg body weight respectively, they exhibited a negative nitrogen balance and their plasma urea nitrogen decreased from 8-11 mg to 2-3 mg per 100 ml. Their plasma-free amino-acid pattern also showed some abnormalities, with the branched-chain amino-acids, valine, isoleucine and leucine values decreasing, indicating some degree of protein depletion (Huang, 1982). This finding confirms that sweet potato protein alone cannot meet adequately the nutritional requirements of a growing child, but appears to be more promising in the case of adults. In an attempt to improve the diets of the people of Taiwan, Yang (1982) found that when 13 percent of sweet potato was substituted equicalorically for rice in the Taiwanese diet, the nitrogen balance was improved owing to complementarity of the proteins. The same replacement was found to prolong the longevity of tested male and female rats. Thus, if it can be produced at a competitive price, sweet potato can provide a supplementary staple for rice, wheat flour and other cereals.

Food containing about 5 percent of total energy provided by utilizable, balanced protein can sustain health if it can be eaten in sufficient quantities to meet energy requirements. It is therefore important to review the factors affecting the protein content of root crops. If varieties with a high protein content and good carbohydrate digestibility could be developed these could be used in the formulation and production of supplementary weaning foods. Experimental production of weaning foods containing potato has been reported by Abrahamsson (1978). Breeding programmes for improved protein, vitamin or mineral content in food crops should also include consumer preference studies, to ensure acceptance of the improved varieties at producer level.

LIPIDS

All the root crops exhibit a very low lipid content. These are mainly structural lipids of the cell membrane which enhance cellular integrity, offer resistance to bruising and help to reduce enzymic browning (Mondy and Mueller, 1977) and are of limited nutritional importance. The content ranges from 0.12 percent in banana to about 2.7 percent in sweet potato. The lipid may probably contribute to the palatability of the root crops. Most of the lipid consists of equal amounts of unsaturated fatty acids, linoleic and linolenic acids and the saturated fatty acids, stearic acid and palmitic acid. In dehydrated products such as dehydrated potato or instant potato, the high percentage of unsaturated fatty acids in the lipid fraction may accelerate rancidity and auto-oxidation, thereby producing off-flavours and odour. The low fat content of plantain, coupled with its high starch content, makes it an ideal food for geriatric patients. Banana is the only raw fruit permitted for people suffering from gastric ulcer, and is also recommended for infantile diarrhea. Banana is also used as a source of carbohydrate in coeliac disease and in the relief of colitis.

VITAMINS

Since roots and tubers are very low in lipid they are not in themselves rich sources of fat-soluble vitamins. However, provitamin A is present as the pigment beta-carotene in the leaves of root crops, some of which are edible. Most roots and tubers contain only negligible amounts of beta-carotenes with the exception of selected varieties of sweet potato. Deep coloured varieties are richer in carotenes than white cultivars. In the orange variety ''Goldrush'', the pigment is made up of about 90 percent beta-carotene and in "Centennial" the corresponding figure is 88 percent. This is one of the nutritional advantages of sweet potato because sufficient and regular ingestion of sweet potato leaves, together with the tubers of high beta-carotene varieties can meet the consumer's daily requirement of vitamin A, and hence prevent the dreadful disease of xerophthalmia, which is responsible for nutritional blindness in many sub-Saharan countries and in Asia. The dessert type of sweet potato is even higher in beta-carotene and it has been estimated that an intake of 13 g/day will be sufficient to meet the vitamin A requirement. Similarly some varieties of yam are highly coloured, especially *D. cayenensis*, called yellow yam. The colour of yellow yam is also because of carotenoids, consisting mainly of beta-carotene in quantities of 0.14-1.4 mg per 100 g (Murtin and Ruberté, 1972) and other carotenoids which have no nutritional significance (Martin *et al.*, 1974b). Some Pacific Island varieties of yam contain up to 6 mg per 100 g (Coursey, 1967) of carotene; cocoyam also has a generous amount. Other sources of beta-carotene include the deep orange varieties of banana. The concentration, however, decreases from 1.04 mg per 100 g when green (unripe) to 0.66 mg when ripe (Asenjo and Porrata, 1956). Plantain contains very little beta-carotene.

Potato has no vitamin A activity. There is some report of the occurrence of some vitamin E, up to 4 mg per 100 g in sweet potato.

Vitamin C occurs in appreciable amounts in several root crops. The level may be reduced during cooking unless skins and cooking water are utilized. Root crops, if correctly prepared, can make a significant contribution to the vitamin C content of the diet. Banana contains about 10-25 mg of vitamin C per 100 g, though figures as high as 50 mg have been quoted in some varieties. The quantity is the same whether it is ripe or unripe. Yam contains 6-10 mg of vitamin C per 100 g and up to 21 mg in some cases. The vitamin C content of potato is very similar to those of sweet potato, cassava and plantain, but the concentration varies with the species, location, crop year, maturity at harvest, soil, nitrogen and phosphate fertilizers (Augustin *et al.*, 1975). One hundred grams of potato boiled with the skin is sufficient to provide about 80 percent of the vitamin C requirement of a child and 50 percent of that for an adult. According to the 1983 Nutritional Food Survey Committee, potato was a principal source of vitamin C in British diets, providing 19.4 percent of the total requirement. McCay *et al.* (1975) estimated that in the United States of America potato provided as much vitamin C (20 percent) as did fruits (18 percent).

Most of the root crops contain small amounts of the vitamin B group, sufficient to supplement normal dietary sources. The B-group of vitamins acts as a co-factor in enzyme systems involved in the oxidation of food and the production of energy. These vitamins are found mainly in cereals, milk and milk products, meat and green vegetables, including the leaves of roots and tubers. For every 1 000 kcal of carbohydrate ingested about 0.4 mg of vitamin B_1 (thiamine) is needed for proper digestion. Sweet potato contains about double this required amount of vitamin B_1 (0.8-1.0 mg/1 000 kcals). Villareal (1982) has estimated that a hectare of land planted with sweet potato will provide about eight times as much vitamin B_1 (thiamin) and 11 times as much vitamin B_2 (riboflavin) as a hectare planted with rice (see table 4.12). Similarly it has been estimated by the Nutrition Food Survey Committee (1983) that in the United Kingdom potato supplied 8.7 percent of the riboflavin, 10.6 percent of the niacin (vitamin B_3), 12 percent of the folic acid, 28 percent of the pyridoxine (vitamin B_6) and 11 percent of the panthothenic acid (Finglas and Faulks, 1985).

TABLE 4.12

Number of persons a hectare of crop can support per day in terms of different nutrients

Crop	Calories	Calcium	Iron	Vitamin A	Thiamin	Riboflavin	Vitamin C
Rice	61	2	33	0	18	9	0
Maize	27	1	9	25	42	24	480
Sweet potato	138	138	405	991	140	106	1 370
roots	122	85	105	324	100	40	1 050
leaves	15	53	300	667	40	66	320
Taro	55	86	178	770	120	61	660
corms	45	28	71	0	107	24	180
leaves	6	40	65	747	10	33	433
petiole	3	16	40	23	1	3	46
Cabbage	41	178	194	50	92	74	3 441
Mungo	29	17	78	4	60	20	27
pod	42	159	150	347	158	168	1 008
dry bean	63	18	193	0	129	61	0
Soybean *(dry)*	33	41	168	0	40	16	trace
Soybean *(green)*	36	87	194	6	1 257	614	251
Mango	1	0	501	18	1	1	279
Tomato	16	26	116	257	58	38	845
Banana	2	110	2	1	0	2	237

Source: Villareal, 1970

MINERALS

Potassium is the major mineral in most root crops while sodium tends to be low. This makes some root crops particularly valuable in the diet of patients with high blood pressure, who have to restrict their sodium intake. In such cases the high potassium to sodium ratio may be an additional benefit (Meneely and Battarblee, 1976). However, high potassium foods are usually omitted in the diet of people with renal failure (McCay *et al.*, 1975). As root crops are low in phytic acid relative to cereals, those minerals liable to inactivation by dietary phytic acid are more available than in cereals. This is especially important for iron, which has been found to be 100 percent available in banana (Marriott and Lancaster, 1983). In addition the high vitamin C concentration in some root crops may help to render soluble the iron and make it more available than in cereals and other vegetable foods. In the United Kingdom the iron supply from potato ranks third of all individual food sources, accounting for up to 7 percent of the total household dietary iron intake. True *et al.* (1978) found that 150 g of potato will supply 2.3 to 19.3 percent of the dietary requirement for iron recommended by the Food and Nutrition Board of the National Research Council of America. However there is some doubt about the availability of calcium and phosphorus in cocoyam owing to the oxalate content.

An important, often unrecognized, mineral contribution that potato can make is in the appreciable amount of iodine it contains. This could be significant in goitrous areas of Africa and Asia where iodine intake is low or marginal. Since over 96 percent of the zinc in potato is available, again due to low levels of phytate, potato can also significantly contribute this mineral. Yam can supply a substantial portion of the manganese and phosphorus requirement of adults and to a lesser extent the copper and magnesium. As indicated in Table 4.12, a hectare of sweet potato will provide the calcium requirement for 60 times as many people and 12 times the requirement of iron as the same area of land planted with rice.

ROOT CROP LEAVES

Apart from the yellow variety of sweet potato, which contains a high amount of beta-carotenes (up to 30 mg retinol equivalent percent) most of the other root crops contain only negligible amounts. However, their leaves contain a substantial amount of beta-carotenes that could contribute significantly to the daily requirement of vitamin A, especially for children, thereby helping to eradicate the ocular diseases affecting about six to eight million children from Asia, Africa and Latin America. Dietary retinol obtained from the consumption of animal foods is relatively expensive and contributes about 14 and 20 percent to the vitamin A intake of people in Asia and Africa respectively. Beta-carotenes from leaves such as sweet potato or cassava, which contain about 800 mg/100 g and which is about the same as beef liver, contribute as much as 86 percent in Asia and 80 percent in Africa.

The quantity of root crop leaves required to meet the average daily requirement of retinol differs considerably, with cassava requiring only 50 g, dark green vegetable leaves 73 g, sweet potato leaves 78 g and taro leaves 133 g.

Cassava leaves have a crude protein content of 20-35 percent on a dry weight basis. The quality of the leaf protein is generally good though it is deficient in methionine. Cassava leaves are low in crude fibre and relatively high in calcium and phosphorus. Cassava varieties in which the tuber contains cyanogenic glycosides usually show a similar content in their leaves.

5. Methods of cooking and processing

Like many other foods, roots and tubers are rarely eaten raw. They normally undergo some form of processing and cooking before consumption. The methods of processing and cooking range from simple boiling to elaborate fermentation, drying and grinding to make flour, depending on the varieties of roots and tubers.

The basic purpose of these methods is to make roots and tubers and their products more palatable and digestible and to make them safe for human consumption. Processing also extends the storage life of roots and tubers, which are often highly perishable in their fresh condition. Processing also provides a variety of products which are more convenient to cook, prepare and consume than the original raw materials.

Women play a very active role in all the stages involved in the production and processing of root crops. Assessment from five states in Nigeria indicated that in cassava production women on average complete 34 percent of the field preparation and 77 percent of the planting of cassava, 86 percent of the weeding and 77 percent of the harvesting. The post-harvest activities of processing, storage and marketing, are undertaken mainly by women, though recent studies indicate that men are beginning to take an interest in the processing of root crops through the purchasing and management of electrically operated grinding machines.

CASSAVA

Although raw sweet cassava is occasionally eaten in the Congo region, Tanzania and West Africa, cassava is not generally consumed raw. A large variety of processing techniques have been developed in different parts of the world resulting in a wide variety of products. Those techniques serve not only to render the root palatable, and in many cases storable, but they also have the effect of eliminating or reducing cyanide (HCN) content to acceptable levels. Many processes such as soaking and fermenting have been designed specifically to detoxify the root. Others, such as boiling and roasting are designed to make cassava products more palatable. The degree of reduction of cyanide in the final product varies greatly with the type of processing techniques used. Many of the complex techniques used around the world today originated in South America and were introduced to the other regions with the cassava plant, or in some cases at a later date. Other processes have been developed independently in the producing countries.

Roasting, boiling, frying

In Latin America roasting is the simplest technique, but it is not commonly practised. It is used only when no cooking utensil is available. The whole roots are buried in hot ashes or placed in front of the fire until cooked through.

Sweet cassava roots are more often prepared by boiling and are eaten either hot or cold or sometimes mashed. These are general methods all over the world. In Latin America, a soup or stew called *cancocho* or *cocido* is prepared by boiling cassava roots with vegetables. The technique of deep-frying cassava in fat is thought to have been introduced by Europeans. In Uganda, the roots are peeled, washed, wrapped in banana leaves and steamed in a pan (Goode, 1974). Roasting sweet cassava in ashes is widely practised in Africa. In South Africa, bitter varieties are also roasted but are first peeled and rubbed with tobacco. In Zambia roots are often soaked before roasting. Fried cassava is prepared after peeling, by washing, slicing and then frying in oil.

Grating, pounding, baking, or boiling

In Latin America, cassava roots are grated on spiny palm trunks or pounded into a pulp. The pulp is then squeezed by hand and cooked in a variety of ways. Several groups shape the pulp into pies or cakes which are then baked in hot ashes, sometimes being wrapped in a protective covering of leaves before baking. Some groups, such as the Nambicuara, sun dry the pulp balls, wrap them in leaves and place them in a basket or bury them in the ground, to be used at times of food shortage. After several months, they retrieve the fermented balls and cook them by baking in hot ashes. The cassava pulp is boiled either by dropping the pie or balls into the boiling water or by stirring the pulp into water to form a sort of porridge. Porridge is sometimes made as a preliminary step in the preparation of flour. The pulp is boiled and skimmed off with a plated spoon, strained through a mat of thin sticks and finally roasted in a pan to make flour.

Steaming and fermenting peujeum. A traditional product prepared in Java is *peujeum* (Stanton and Wallbridge, 1969). The peeled cassava roots are steamed until tender, allowed to cool and then dusted with finely powdered *ragi,* a rice flour starter culture flavoured with spices. The cassava mash mixed with *ragi* is wrapped in banana leaves in an earthenware pot and left for one or two days to ferment. The *peujeum* has a refreshing acidic and slightly alcoholic flavour and is either eaten immediately or baked.

Sun drying and pounding or grinding into flour

Cassava roots, which may be soaked in water first, are sun dried and pounded into a flour. This seems to be a general method everywhere.

In the preparation of *fuku* in Zaire the dried roots are pounded with partially fermented corn, the quantity depending on the season. The resulting flour is roasted on a flat tray to stop further fermentation of the mixture, which was initiated by the fermenting maize. The flour is eaten as a gruel prepared in boiling water. Cassava flour is the basis of several other foods. In the preparation of *nsua* the flour is mixed with water and filtered through a jute bag. After removal of the water the paste is wrapped in a leaf and eaten raw. *Ntinga* is prepared in a similar way except that a portion of the paste is boiled in water and mixed with the remaining uncooked paste. The mixture is wrapped up in a leaf and cooked again.

Grating, pressing and roasting or baking to make flour or bread

These methods are widely used to prepare cassava flour or cassava bread in tropical America. Details vary from one group to another but the methods fall into two main groups depending on whether or not the roots are given a preliminary soaking:

Unsoaked roots. The process is very laborious and takes two or more days. The freshly dug roots are first washed to remove excess soil and then peeled. The tubers are then reduced to a pulp, normally by grating, but sometimes by crushing in a mortar or between stones. The pulp is squeezed with a variety of devices to expel the liquid. The moist pulp is left overnight in containers. Next day it is rubbed through a sieve to remove any coarse fibres. The pulp is then cooked in one of two ways depending on whether bread or flour is needed.

To prepare bread, the cassava pulp is placed on a hot clay or stone griddle, pressed down into a thin layer and toasted on each side. The large, flat circular cakes are known as cassava bread, cassava *casabe, beigu* or *couac* depending on the locality. When fresh, the bread is soft inside and some people prefer to prepare it daily. More commonly, cassava bread is sun dried for several days during which time it hardens through. It can be stored in this form for several months. Cassava bread is normally eaten dipped into gruel or stew, which serves to soften it (Jones, 1959). Other types of bread can be made by adding additional ingredients to the cassava, for example in Brazil a special bread is prepared by adding pounded or grated Brazil nut to the cassava pulp.

To prepare flour, the cassava pulp is stirred continuously while cooking on the griddle in order to prevent lumps forming. The resulting flour also stores well and is variously known as *farinha de mandioca, farinha seca, farinha surruhy, kwak* or *koeak*. It may be eaten dry, mixed with hot or cold water to make a paste or gruel or mixed with other foods. Various modifications and other methods, both simple and complex, are also used.

A traditional Philippine dish based on the cassava root is known as cassava rice or *landang*. Freshly dug roots are peeled and grated, the grated mass is put into jute sacks and pressed between two wooden blocks to squeeze out the juice. It is then put into a winnowing basket and whirled until pellets are formed. At intervals the pellets are sieved and those not passing through are put back and whirled again. The pellets are dried on a mat and then steamed in a coconut shell on a screen mesh placed over a vat of boiling water. The cooked pellets are placed in the winnowing basket and separated by hand. In an alternative process the peeled roots are submerged in fresh, clean water in an earthenware jar or wooden vessel for five to seven days until soft. They are then macerated, the fibres are removed and the remainder is dried and made into pellets as described. The pellets from both methods are dried thoroughly in the sun for three to five days and stored until needed. Cassava rice can be eaten without further cooking.

Soaked roots. In Latin America the cassava tubers, either peeled or unpeeled, are soaked in water usually for three to eight days but sometimes even longer to allow some fermentation to occur. After removal from the water the peels are removed where necessary and the softened roots are either crushed by hand or grated to make a pulp and processed as for *farinha seca*. This method is also used to prepare cassava bread but more often the end product is cassava flour. Many variations of this basic process exist.

In West Africa, after fermentation the cassava is pounded or ground to produce a paste which is consumed or stored depending on the country. In parts of Nigeria, the paste is boiled for 20 minutes and then removed for more pounding. In Cameroon, the wet paste is divided into two portions and wrapped in leaves before cooking. In Mozambique cassava paste is mixed with flavouring ingredients including onion and salt, before being wrapped in leaves and boiled.

The preparation of pastes by pounding cassava is a peculiarly African process not practised in South America. The pastes are consumed in a variety of forms, the best known being *fufu*. The term *fufu* and its variants are widely used in West Africa to refer to a sticky dough or porridge prepared from any pounded starchy root including yam, cocoyam and cassava.

In the preparation of *fufu* the roots are peeled, washed, grated and left to ferment for two to three days. To ferment the cassava the grated mass is either simply left to stand (Doku, 1969) or put into sacks and weighed with stones to squeeze out the juice. The resulting dough is used at once for cooking or it is stored in basins covered with cold water which is changed daily. The resulting product is consumed in different ways in different countries accompanied by stew or soup.

Gari is the most popular cassava product consumed in Africa. To prepare *gari* cassava roots are washed, peeled and grated. The pulp is then placed in cloth bags or sacks made from jute and left to ferment, the fermentation time varying from three to six days. It is the fermentation process that gives *gari* its characteristic sour flavour, which distinguishes it from Brazilian *farinha*. During this stage pressure is applied to squeeze out the cassava juice. The cassava pulp, at about 50 percent water content, is taken out of the sacks and sieved to remove any fibrous material. It is then heated or "garified" in shallow pans and stirred continuously until it becomes light and crisp.

Gari is eaten in a variety of forms. It is sometimes eaten dry or made into a paste. More commonly it is steeped in cold water, thus causing the particles to swell and soften but retaining their granular form. Alternatively, *gari* is mixed with cold water to make a thin gruel which is drunk with milk. A popular way of preparing *gari* in Nigeria is to soak it in boiling water to form a thick paste, *eba*, sometimes known as *fufu*.

Products essentially similar to *gari* are known by various names and are made throughout West Africa with minor variations to the processing. Recently *gari* processing has been mechanized in Nigeria.

A regional standard for Africa for *gari* was adopted by the *Codex Alimentarius Commission* (1986) which classified *gari* into five categories according to grain size and specified their essential composition and quality factors. These include raw material cassava and characteristic colour, taste and odour of *gari* and specification on acidity (not less than 0.6 percent nor more than 1 percent m/m determined as lactic acid). Total hydrocyanic acid (not exceeding 2 mg/kg determined as free HCN), moisture (not exceeding 12 percent m/m), crude fibre (not exceeding 2 percent m/m), ash content (not exceeding 2.75 percent m/m) and should be practically free from extraneous matter. Optionally edible fats or oils and salt may be added and *gari* may be enriched with added vitamins, proteins and other nutrients but addition of food additives was not allowed.

The methods used to process cassava in the South Pacific vary from island to island although boiling or baking the tubers are fairly widespread techniques. In the Solomon Islands the roots are often grated and mixed with coconut or banana as a pudding. In the New Hebrides, cassava is grated, wrapped in banana leaves and baked in an oven.

One method peculiar to the islanders in the South Pacific is the fermentation of tuber roots in pits, a process which prolongs the shelf-life of the product indefinitely. On the island of Mango in Tonga an abandoned pit estimated to be about 100 years old was found to contain food in an edible condition. Traditionally the pit is dug to a depth depending on the size of the family and lined with leaves of coconut, giant swamp taro or banana. The prepared food, which could be cassava, banana, taro or a mixture of all three, is placed in the pit to fill it and covered with more leaves, with rocks or logs placed on top to keep it in place. Fermentation proceeds for four to six weeks, after which the whole or part of the product is removed. Sometimes the fermentation is carried out with the addition of fresh water and sea water. In a modification of this process in Fiji, the unpeeled cassava root is fermented in a basked lowered into a lagoon. When it is required, it is removed, drained and pounded to a dough. The dough is kneaded with previously grated coconut, formed into balls, wrapped in breadfruit leaves and eaten either steamed or boiled. This product keeps for several months. If fresh water is used for the fermentation, the pulp is mixed with sugar or fruit, wrapped in leaves and steamed or boiled. This is known as *bila* and is a favourite food in Fiji. It keeps for several days.

Extraction of starch to prepare *sipipa*, tapioca and pot *bammie*

The juice obtained from grated cassava contains a certain amount of starch which settles out on standing for a few hours. In the Americas the liquid is decanted off, the starch residue is rinsed and then processed. It may be sun dried and eaten raw, or baked into crisp cakes called *sipipa*, which are highly prized as a delicacy by some groups. If it is still wet the starch is heated on a griddle when the starch grains burst and form granules known as tapioca flakes or globules. In Jamaica, starch is obtained by mixing grated cassava roots with water and straining the pulp through a towel. The starch is allowed to settle out for a few hours. The water is decanted off and the starch is either dried briefly, then salted and baked into pot *bammie,* or dried for several days, pounded in a mortar, mixed with flour and cooked into dumplings.

In Asia the traditional methods of extracting starch are similar to those used in tropical America and Africa. The starch in the extracted cassava juice is washed and sun dried on a mat. Moist wet starch is used commercially to produce tapioca. To prepare tapioca the wet starch is gelatinized into globules and sun dried.

In the South Pacific starch is extracted from cassava roots by grating, washing and draining and is then dried in an oven to produce a granular, tapioca-like product.

In Padaids Island the pulp, from which the starch has been extracted, is also used. It is formed into balls of five to six cm in diameter and dried over a fire for about a week. When required for eating the dried cassava is grated again and mixed with coconut milk and water (Massal and Barrau, 1956).

In the Solomon Islands of Anuta and Tikopia, cassava is used to produce a fermented product called *ma manioka* on Anuta and *masi manioka* on Tikopia (Yen, 1978). On Tikopia, the cassava roots are soaked in water for five or more days until soft. They are then peeled, broken up, squeezed and ensiled in pits lined with leaves. On Anuta, where there is no suitable surface water, the roots are packed loosely in pits and left for a few weeks. They are then recovered, peeled and returned to the same pits for a further period. *Ma* is used as an emergency food baked alone or in combination with freshly pounded starchy roots and fruits.

Processing cassava juice - cassava reep, beer

The cassava juice or *yari,* obtained by pressing the grated cassava, is commonly used to prepare sauces and beverages in South America and the West Indies. The *yari* is boiled down to a thick syrupy consistency. The soup is known as cassava reep in the West Indies. Groups inhabiting the area around the headwaters of the Amazon tributaries produce a refreshing sweet-tasting drink by boiling *yari* for several hours. An alcoholic drink may also be prepared by fermenting the cassava juice.

Preparation of beverages from cassava root

In addition to cassava juice, the whole root, the sliced, grated or pounded roots and cassava bread or flour are all used as starting materials for the preparation of beverages. Both non-alcoholic and alcoholic drinks are made.

Non-alcoholic drinks. The roots are peeled, grated, squeezed by hand and cooked. When cold they are masticated for a few minutes and allowed to stand for a short period but not long enough to produce an alcoholic drink. Similar drinks are made from cassava flour or bread.

Alcoholic drinks . The preparation of cassava beers is widespread in tropical America. These are known as either *kashiri* or *chicha.* A number of different methods are used. The most common methods are the following:

Processing without mastication. The drink is usually prepared by fermenting whole cassava roots. The tubers are left for up to a week in a stream for fermentation to occur. They are then removed and mashed. Water is added to the mash which is left to stand for three days before drinking. Other methods of preparation are also used.

Many groups use cassava bread to prepare beverages. In Guyana freshly made cassava bread is dipped into water, placed in a shallow heap in a dark corner of the house and left, covered with leaves, for three to five days while moulds develop. The broken bread is then placed in large earthenware pots and left for a further two to five days. Finally water is added and fermentation takes place to produce a mildly intoxicating beverage. Other methods are used in Brazil and in Suriname to prepare alcoholic drinks from cassava bread.

Processing with mastication. The custom of mastication in the preparation of alcoholic drinks is common in tropical America. The majority of the traditional alcoholic drinks are prepared in this way. Mastication speeds up fermentation because the salivary enzymes initiate the conversion of starch to sugar.

A variety of beverages is prepared from masticated cassava. In the Brazilian tropical forest, thinly sliced and boiled pieces of cassava are squeezed, chewed and left to ferment for one to three days. In the West Indies, a drink known as *paiwari* is prepared by this method. Other fruits, vegetables, maize and sweet potato may also be added as ingredients to the beer.

Beverage making from cassava is not generally practised in Africa. Goode (1974) describes a method of preparation of beer in Uganda. The flour is mixed with water and left to ferment for a week. It is then roasted over a fire and put into a container to which water and yeast are added. After about a week the liquid is drained, sugar is added and the beer is left to ferment for four days. Cassava flour is also used to make beer in South Africa, Southwest Zambia and Angola.

COOKING AND PROCESSING OF YAM

By far the greater part of the world's yam crop is consumed fresh. Traditionally processed yam products are made in most yam-growing areas, usually as a way of utilizing tubers that are not fit for storage.

Usually fresh yam is peeled, boiled and pounded until a sticky elastic dough is produced. This is called pounded yam or yam *fufu*.

The only processed yam product traditionally made at village level is yam flour. Except by the Yoruba people in Nigeria, yam flour is regarded as an inferior substitute for freshly pounded yam because it is often made from damaged tubers. Yam flour is favoured in the Yoruba area where the reconstituted food is known as *amala*. To a limited extent, yam flour is also manufactured in Ghana where it is known as *kokonte*. The nutritional value of yam flour is the same as that of pounded yam.

Preparation of yam flour

The tubers are sliced to a thickness of about 10 mm, more or less, depending on the dryness of the weather. The slices are then parboiled and allowed to cool in the cooking water. The parboiled slices are peeled and dried in the sun to reduce the moisture content.

The dried slices are then ground to flour in a wooden mortar and repeatedly sieved to produce a uniform texture. Today small hand-operated or engine-driven corn mills or flour mills are increasingly used.

Industrial processing

Yams have not been processed to any significant extent commercially. Dehydrated yam flours and yam flakes have been produced by sun drying. The manufacture of fried products from *D. alata* has also been attempted recently. Both chips and French fries have been manufactured. Preservation of yam in brine has been attempted, but with little success.

Since pounded yam has so much prestige and is the most popular way of eating yam, two attempts have been made to commercialize the process. The first was the production of dehydrated pounded yam by drum drying. This product could then be reconstituted without further processing. This production was first attempted in Côte d'Ivoire in the mid-1960s, under the trade name "Foutoupret", by air drying precooked, grated or mashed yam (Coursey, 1967). Onayemi and Potter (1974) used drum drying to produce a flake which can easily be reconstituted into pounded yam by mixing with boiling water. This is the basis of the commercial product called "Poundo" in Nigeria, which was initially successful. To reduce wastage of raw material, peeling is done by using a 10 percent lye at 104°C with varying immersion times depending on the cultivar of yam (Steele and Sammy, 1976). Sulphite is added to prevent enzymic browning.

In the second commercial project a type of food processor resembling a blender was developed. The yam is cooked, fumed and churned in a process equivalent to pounding, to give enough pounded yam for two to four servings. Both projects appeared at first to be very successful, but later people reverted to the manual pounding of yam which gives a characteristic viscosity and firmness that is difficult to simulate mechanically.

Attempts to manufacture fried yam chips, similar to French fried potatoes have been reported from Puerto Rico.

COCOYAM

Cocoyam is used in essentially the same way as yam. It can be eaten boiled, fried or pounded into *fufu,* although it is not considered as prestigious as yam. It has also been made into porridge or pottage, as well as chips and flour. Cocoyam flour has the added advantage that it is highly digestible and so is used for invalids and as an ingredient in baby foods.

Taro is the traditional staple in the Pacific Islands, where it is made into a series of food products similar to those described for cassava. *Poi* is a very popular Hawaiian and Polynesian dish made by pressure cooking the raw tuber, which is then peeled and mashed to a semi-fluid consistency. It is passed through a series of strainers, the final one having openings of about 0.5 mm diameter. It is then bagged and sold, or else stored at room temperature where it undergoes lactic acid fermentation. Coconut products can be added to the fermented *poi* before consumption.

In Nigeria cocoyam is grated, mixed with condiments and wrapped in leaves. It is steamed for about 30 minutes and served with sauce. Popularly known as *ikokore,* it is very common in the western Nigeria. A modification is available in Cameroon where cocoyam is made into balls and cooked with additional ingredients. This is known as *epankoko.*

BANANA AND PLANTAIN

One advantage of banana is that the dessert varieties can be eaten raw without any further processing. In many parts of Africa cooking banana is prepared by boiling or steaming, mashing, baking, drying or pounding to *fufu*. In Cameroon, green banana is boiled and served in a sauce of palm oil with fish, cooked meat, green beans, haricot beans and seasonings. In Uganda, where it is the staple, it is boiled with other ingredients including beans. Ghee is added together with pepper, salt and onions. This dish is called *akatogo*. *Omuwumbo* is prepared by wrapping the pulp in banana leaves and steaming it for about an hour. It is then pressed in the hands to a firm mass and eaten. The green form of banana is dried and stored. Known as *mutere,* it may be used for cooking after grinding into flour (Goode, 1974), but it is mainly used as a famine reserve. The same procedure is used in Gabon, in Cameroon, in South and Central America and in the West Indies (Fawcet, 1921).

A soup called *sancocho* is made in Colombia by boiling slices of green banana with cassava and other vegetables, while in the West Indies boiled green banana is served with salted fish or meat.

Mention has already been made of the fermentation of banana in pits in the Pacific. The fermented product is formed into loaves and baked.

Known as *masi,* it keeps for over a year while buried in the pit, and baked *masi* stored in air-tight baskets in a deep hole may last for generations (Cox, 1980). The starch pseudo-stem and corm of the false banana, or *ensete,* is prepared by similar methods in Ethiopia. The fermented product, called *kocho* is used to prepare a flat, baked bread. Ripe bananas are preserved by sun drying. Known as banana figs, they are eaten as a sweetmeats. This product keeps for months or even years.

In West Africa bananas are parboiled before drying. Oven drying is practised in Polynesia. The dried product is then bound tightly in leaves and stored until it is needed (Massal and Barrau, 1956). A similar technique is practised in India.

In Burundi where banana occupies about 25 percent of the arable land, it is mainly used for the production of beer. It has been estimated that local beer is consumed at a rate of 1.2 l/caput/day. Making beer from banana is common in East Africa. Green banana is buried in pits covered with leaves to ripen for about a week, at which stage it also starts to ferment. The peels are removed, the pulp is mixed with grass in a trough and the juice is squeezed out. The residue is washed and added to the bulk of the juice. Roasted sorghum flour or millet is added and the mass is fermented for one to two days, covered with fresh banana leaves. In a modification of the process, honey is added to the fermented banana pulp.

SWEET POTATO

Sweet potato can also be eaten boiled, fried or roasted. When sliced, dried in the sun and ground, it gives a flour that remains in good condition for a long time. In Indonesia sweet potato is soaked in salt water for about an hour to inhibit microbial growth before drying. The flour is used as a dough conditioner in bread manufacturing and as a stabilizer in the ice-cream industry.

Sweet potato has been processed into chips (crisps) in much the same way as potato and the product is now popular in Asia. The sugar-coated chips are popular in China, the salted variety is popular in the United States of America, while those spiced with cayenne pepper and citric acid have been tested in Bangladesh with favourable results (Kay, 1985).

Starch is produced from sweet potato in much the same way as from the other starchy roots except that the solution is kept alkaline (pH 8.6) by using lime, which helps to flocculate impurities and dissolve the pigments. The starch shows properties intermediate between potato starch and corn/ cassava starch in terms of viscosity and other characteristics. In Japan about 90 percent of the starch produced from sweet potato is used in the manufacture of starch syrup, glucose and isomerized glucose syrup, lactic acid beverages, bread and other food manufacturing industries.

In Japan, sweet potato starch is also used for the production of distilled spirits called *shochu* (Sakamoto and Bouwkamp, 1985). The process is very similar to that of whisky production except that the *koji*, equivalent to the malt starter in whisky production, is obtained by inoculating steamed rice soaked in water overnight with *Aspergillus kawachii* for two days at 35 to 37°C. The *koji* is mixed with starch water and yeast to promote saccharification and fermentation. The filtrate is finally distilled. The yield is about 800 l from 1 tonne of sweet potato.

POTATO

Like other root crops potato can be eaten boiled, fried or roasted. Since it is essentially a temperate root crop its use has been extensively commercialized. French fries and potato crisps are very popular snacks in the United States of America and else where. Unlike cereal starches the starch from potato sets rapidly at high temperatures and has a high hot-paste viscosity which makes it preferable for the manufacture of adhesives. It also finds applications in the textile industry, the food industry and in the production of alcohol and glucose. Most of these processes are mechanized and highly efficient. For short domestic storage, peeled potatoes are immersed in a solution of sodium metabisulphite to inhibit discolouration by enzymic action. They can then be refrigerated for several days before cooking and consumption.

The preparation of crisps is very similar to that of French fries except that the former are cut into very thin slices while the latter are cut into rods. The flour produced from potato is incorporated into bread, and is used as a thickener in dehydrated soups, gravies, sauces and baby foods. Dehydrated potato dice are ingredients in processed foods including canned meat, meat stew, frozen meat pies and salads.

Woolfe (1987) has given a detailed description of the processing of bitter potato in the Andes, especially of those varieties that contain toxic alkaloids. In the preparation of *chuño blanco* the potato is evenly spread on the ground on a very frosty night. If, on the following day it is not well frozen, it is exposed for another night. The successive freezing and thawing separates the tuber cells and destroys the differential permeability of the cell membrane, thereby allowing the cell sap to diffuse out from the cell into the intracellular spaces (Treadway *et al.*, 1955). In this way by trampling the thawing tubers the liquid is squeezed out and the skins are removed. The residue is recovered and immersed in a running stream for one to three weeks, to remove toxins. After draining it is spread in the sun to dry. During the drying period a white crust forms on the tubers, which is the origin of the name of the food. *Chuño blanco* forms the basis of soup and stews. It is a delicacy among the inhabitants of the high Andean areas of Peru and Bolivia especially when served steaming hot with cheese.

The preparation of *chuño negro* is very similar to that of *chuño blanco* except that during trampling the skin is not removed, the soaking process is omitted, and the residue is simply sun dried after trampling. The product is dark brown-black in colour, and hence the name. It is usually soaked in water for one or two days before being cooked in order to remove any residual bitter flavours.

A more prestigious potato preparation that is popular in large cities and in Peru is *papa seca*. The potato is boiled, peeled, sliced and sun dried and then ground into a fine flour. The flour is normally used for a special dish called *carapulca* which is prepared with meat, tomato, onions and garlic, but it may also be made into soup.

These traditional techniques are particularly important for processing the bitter varieties of potato with a high alkaloid content, which would otherwise have been too toxic for food use. Christiansen (1977) found that the level of glycoalkaloids could be reduced, from 30 mg/100 g in the fresh potato, to about 4 mg in *chuño blanco* and 16 mg in *chuño negro*. In the Andean highlands, where frost, storm or drought can lead to destruction of crops, irregular yields and food shortages, it is essential to cultivate some frost-resistant bitter varieties of potato that can be processed into reserve food from year to year.

A good review of simple technologies for root crop processing is provided by the United Nations Development Fund for Women publication, *Root crop processing*, 1989.

6. Effect of processing on nutritional value

Root crops are not easily digested in their natural state and should be cooked before they are eaten. Cooking improves their digestibility, promotes palatability and improves their keeping quality as well as making the roots safer to eat. The heat used during cooking can be dry heat as in baking in an oven or over an open fire, or wet heat as when boiling, steaming or frying. Heat helps to sterilize the food by killing harmful bacteria and other micro-organisms, and it increases the availability of nutrients. Proteins are denatured by heat. In this form they are more easily digested by proteolytic enzymes; cellulosic cell walls that cannot be broken down by monogastric animals like man are broken down, and some anti-nutritional factors such as enzyme inhibitors are inactivated. However, processing may reduce the nutritional value of some root crops as a result of losses and changes in major nutrients, including proteins, carbohydrates, minerals and vitamins.

Nutrients may be lost during cooking in two ways. First, by degradation, which can occur by destruction or by other chemical changes such as oxidation, and secondly by leaching into the cooking medium. Vitamins are susceptible to both processes while minerals are affected only by leaching. Free amino-acids could also be leached or may react with sugars to form complexes. Starches may be hydrolysed to sugars. The percentage loss will depend partly on the cooking temperature and on whether the food is prepared by boiling, baking or roasting. Baking losses may appear deceptively low if expressed on a fresh weight basis, due to the concentration of nutrients by loss of water. However, less damage is done by baking than by canning or drum drying (Purcell and Walter, 1982).

The first step in processing any root crop is usually peeling. This may remove nutrients if it is not done carefully. Cooking losses can be reduced by retaining the skin to minimize leaching and to protect the nutrients. It is sometimes advisable to peel after boiling, and to make use of the cooking water in order to conserve water-soluble nutrients.

Vitamin C is the most thermolabile vitamin and is also easily leached into cooking water or canning syrup. Elkins (1979) reported complete retention of vitamin C in freshly canned sweet potato but the vitamin content dropped to 60 percent of its original value after storage for 18 months. The concentration of the canning syrup did not affect vitamin retention (Arthur and McLemore, 1957). Air drying of thin slices of sweet potato leads to only slight losses of vitamin C.

Boiling may result in a 20 to 30 percent loss of vitamin C from unpeeled roots and tubers as shown in Table 6.1. If peeled before boiling the loss may be much higher, up to 40 percent. Swaminathan and Gangwar (1961) estimated that 10 to 21 percent of the loss is due to leaching into the cooking water and the rest to destruction by heat. Baking losses of vitamin C in unpeeled potato are about the same as in boiling but roasting results in higher losses, while making into crisps seems to be slightly better in terms of vitamin retention. Frying results in the loss of 50 to 56 percent compared to 20-28 percent on boiling unpeeled (Roy-Choudhuri *et al.*, 1963). Streghtoff *et al.* (1946) reported a 28 percent loss during baking and only a 13 percent loss when boiled after peeling. The difference may be that the higher temperature of baking leads to greater destruction of the vitamin. As much as 95 percent of the vitamin C is retained when yam is cooked with the skin on but this is reduced to 65 percent if it is cooked after peeling; 93 percent is retained on frying and 85 percent on roasting. (Coursey and Aidoo, 1966).

Up to 40 to 60 percent of the vitamin C content of potato can be lost on storage (Sweeny *et al.*, 1969; Augustin *et al.*, 1978; Faulks *et al.*, 1982) depending on the temperature. Storage for 30 weeks at 5° or 10°C resulted in a loss of 72 percent and 78 percent respectively (Yamaguchi *et al.*, 1960); and 49 percent loss at 8.5 months (Roine *et al.*, 1955). On the other hand storing for 12 weeks at a tropical humid temperature of 16° or 28°C and 55 percent and 60 percent relative humidity, respectively, resulted in some sprouting and softening of the potato, followed by an increase in the vitamin C content from the initial value of 8.2 mg to 10.1 mg and 10.5 mg per 100 g

TABLE 6.1

Composition of potato, cassava and plantain cooked by different methods (per 100 g)

Cooking method	Energy (kJ)	Energy (kcal)	Moisture (%)	Crude protein (g)	Fat (g)	Total carbo-hydrate (g)	Dietary fibre (crude fibre) (g)	Ash (g)	Ca (mg)	P (mg)	Fe (mg)	Thiamin (mg)	Niacin (mg)	Ribo-flavin (mg)	Pyri-doxine (mg)	Total folic acid (µg)	Ascorbic acid (mg)
Uncooked	335	80	78.0	2.1	0.1	18.5	1.7[1] (0.5)	1.0	9	50	0.8	0.10 (0.2)[1]	1.5 (0.6)[1]	0.04 (0.02)[1]	0.25	14 (35)[1]	20
Boiled in skin,[2] flesh only	318	76	79.8	2.1	0.1	18.5	0.5	0.9	7	53	0.6	0.09	1.5	0.03	-	-	12-16[3]
Boiled peeled	301	72	81.4	1.7	0.1	16.8	1.6[1] (0.6)	0.7	6	38	0.5	0.08 (0.2)[1]	1.2 (0.5)[1]	0.03 (0.01)[1]	0.18	10 (30)[1]	4-14[5]
Baked in skin,[6] flesh only	414	99	73.3	2.5	0.1	22.9	1.9[1] (1.2)	1.2	10	60	0.8	0.10 (0.2)[1]	1.8 (0.6)[1]	0.04 (0.01)[1]	0.18	10 (25)[1]	12-16[3]
Mashed, with milk and margarine[6]	444	106	78.4	1.8	4.7	15.2	(0.7)	1.5	18	40	0.4	0.08	1.1	0.04	0.18	(10)[7]	4-12[5]
Roasted, in shallow fat, flesh only[1]	657	157	64.3	2.8	4.8	27.3	2.7[1]	-	10	53	0.7	0.10 (0.2)[1]	1.9 (0.6)[1]	0.04 (0.02)[1]	0.18	7[5] (35)[1]	5-16[5]
French fried, chips[6]	1 165	264	45.9	4.1	12.1	36.7	3.3[1] (1.0)	1.8	15	92	1.1	0.12 (0.2)[1]	2.6 (0.6)[1]	0.06 (0.02)[1]	0.18	(10)[7] (35)[1]	5-16[5]

(cont.)

TABLE 6.1 *(cont.)*

Composition of potato, cassava and plantain cooked by different methods (per 100 g)

Cooking method	Energy (kJ)	Energy (kcal)	Moisture (%)	Crude protein (g)	Fat (g)	Total carbo-hydrate (g)	Dietary fibre (crude fibre) (g)	Ash (g)	Ca (mg)	P (mg)	Fe (mg)	Thiamin (mg)	Niacin (mg)	Ribo-flavin (mg)	Pyri-doxine (mg)	Total folic acid (µg)	Ascorbic acid (mg)
Chips, crisps[6]	2 305	551	2.3	5.8	37.9	49.7	11.9[5]	3.1	39	135	2.0	0.20	5.5	0.07	0.89	20[5]	17[5]
Canned solids, only[5]	222	53	84.2	1.2	0.1	12.6	2.5	-	11	31	0.7	0.04	0.7	0.03	0.16	11	13
Cassava, boiled	519	124	68.5	0.9	0.1	29.9	0.4	0.6	-	-	-	-	-	-	-	-	26
Plantain, green boiled	393	94	74.5	1.1	0.2	23.8	6.4	0.6	9	32	1.2	0.04	0.6	0.06	-	-	12
Cassava, raw	607	145	62.6	1.1	0.3	35.2	5.2	0.9	38	41	1.0	0.06	0.6	0.04	-	-	36

[1]Finglas & Faulks (1985).
[2]Watt & Merril (1975).
[3]Calculated as by Paul & Southgate (1978).
[4]Average of figures from Paul & Southgate (1978), Watt & Merrill (1975), Wu Leung *et al.*, (1978).
[5]Paul & Southgate (1978).
[6]Average of figures from Paul & Southgate (1978). Watt & Merrill (1975).
[7]Estimated values from Paul & Southgate (1978).
Source: Woolfe, 1987.

respectively. This indicates that storage losses of vitamin C from potato are less in humid tropical conditions than in dry temperate conditions (Linnemann *et al.,* 1985).

Vitamin A is fat-soluble and thermo-stable so it will not normally be degraded by cooking. During studies on the canning of sweet potato, Arthur and McLemore (1957) found no effect on the vitamin A content of the product due to syrup concentration, 0 to 35 percent sucrose, to cooking time, 50 to 90 minutes, or to the peeling conditions. However, Elkins (1979) reported about 14 percent loss of vitamin A activity after processing sweet potato but no additional loss over 18 months, while other investigators have reported a 20 to 25 percent loss of vitamin A activity on cooking. This is probably because of the destruction of the beta-carotene. The main reaction that could take place during the canning of sweet potato is the isomerization of beta-carotene, to neo-beta-carotene leading to a reduction in the vitamin A activity from 95 to 91 percent. The loss will be greater with increasing temperature (Panalaks and Murray, 1970). Losses of carotene and the development of off-flavours occur when sweet potato is stored at an ambient oxygen concentration at which antioxidants are not effective. About 20 to 40 percent of the carotene could be destroyed in the first 30 days from autoxidation. (Deobald and McLemore, 1964). At the same time autoxidation of the lipids, which are highly unsaturated, may occur leading to the development of the off-flavours.

Some of the reported losses in the B-group of vitamins are inconsistent because of differences in the heat lability of the vitamins. Thiamine is thermolabile, but boiling unpeeled potatoes reduced their thiamine content by only 23 percent, drying unpeeled potato led to a loss of only 20 percent while frying after peeling gave a 55 to 65 percent loss (Hentschel, 1969). Riboflavin and niacin are heat-stable and so there is complete retention of these nutrients on boiling, roasting or frying potatoes, although some leaching losses may occur (Finglas and Faulks, 1985). The effect of cooking on the food value of boiled, steamed and baked taro (cocoyam) is shown in Table 6.2. In the case of pyridoxine there is a 98 percent retention of this vitamin on boiling potatoes, the loss being higher with peeled than unpeeled

TABLE 6.2

Effect of cooking on composition of taro (cocoyam)

(results calculated on fresh weight basis)[1]

	Analysis of control (g /kg^{-1})		Differences		
			Boiled-control	Steamed-control	Baked-control
Moisture	655	(10.0)[2]	44.0**	20.0*	-75.0**
Ash	7.6	(0.9)	-0.7*	0.1	0.5
Starch	278	(12.0)	32	29	11
Dietary fibre	12.2	(1.4)	8.2**	7.9**	7.7**
Sugars[2]					
Fructose	1.0	(0.6)	-0.2	-0.1	-0.2
Glucose	0.6	(0.2)	-0.1	-0.1	-0.1
Sucrose	9.4	(1.6)	-0.8	-1.1	-1.3
Maltose	1.0	(0.3)	-0.2	-0.1	-0.1
Minerals mg/kg^{-1}					
Ca	160	(30)	10	6.2	-9.0
P	330	(50)	11	41	45
Mg	320	(40)	-5.8	17	2.6
Na	34	(3.0)	9.3	9.5	-2.3
K	3 280	(360)	-410.0*	18	-60
S	54	(7.0)	-1.2	3.3	4
Zn	4.7	(0.5)	0.2	0.5	0.8
Mn	1.4	(0.5)	0.2*	0.2*	0.3
Al	3.1	(1.3)	0.9	1.1	-1.4*
B	0.9	(0.4)	-0.2	-0.1	-0.1

[1]Results from five corms of cultivar Samoa were averaged, standard deviations in parentheses; differences marked with one asterisk are significant at $P<0.05$, those with two asterisks at $P<0.01$. Other results not included in the table are protein 9.6 (1.5), fat 0.5 (0.3), raffinose 0.3 (0.1) g/kg^{-1}; Fe 7.9 (1.8), Cu 2.0 (0.7) mg/kg^{-1}.
[2]The value of 655 was the moisture content at harvest in Fiji; the moisture content before cooking in Canberra was 582 (17) g/kg^{-1}.
Source: Bradbury & Holloway, 1988.

potatoes (Augustin *et al.*, 1978). However, no loss was reported on baking, roasting or frying, probably due to the concentration of nutrient through loss of water (Finglas and Faulks, 1985). Complete retention of thiamine and nicotinic acid in canned sweet potato has been reported, even after storage for 18 months (Elkins, 1979).

Storage has variable effects on different members of the vitamin B group. In potatoes stored at 5° or 10°C, 30 to 50 percent of their thiamine content is lost after six or seven months. There was a significant increase in pyridoxine level, 154 percent and 86 percent respectively for two varieties of potato kept for six months at 4.5°C (Page and Hanning, 1963).

Raw potato starch is undigestible but digestibility increases with cooking time to 75 percent after 15 minutes and to 90 percent after 40 minutes (Hellendoorn *et al.*, 1975). When the whole tuber is baked, as with sweet potato, virtually all the starch is hydrolysed to dextrin and sugars, mainly maltose. The concentration of reducing sugars is low, probably because of the Maillard reaction with lysine.

Baking may decrease the amount of pectin in roots and the degree of esterification, thereby decreasing their dietary fibre content, but this is not nutritionally significant.

The major change in amino-acids that occurs on cooking is the Maillard reaction which makes lysine unavailable, thereby reducing the nutritive value of the roots. Loss of free amino-acids also takes place through leaching (Meredith and Dull, 1979). When sweet potato was canned in 30 percent sucrose or water, the concentrations of essential amino-acids as a percentage of the original were 70 and 58 percent respectively, aromatic amino-acids 69 and 48 percent and sulphur amino-acids 86 and 60 percent respectively. Purcell and Walter (1982) noted a significant reduction in the lysine and methionine content of sweet potato on canning, which may probably be due partly to leaching.

Boiling does not appreciably reduce the total nitrogen content of potato except for some loss owing to peeling. There is a 0.8 percent loss in the boiled, unpeeled tuber compared to a loss of 6.5 percent in the peeled tuber (Herrera, 1979). Nitrogen loss on roasting is also very small, apart from loss of lysine, with losses being greater in frying than baking.

Minerals are usually lost through being leached into syrup during canning, most especially with potassium, calcium and magnesium (Lopez *et al.* 1980); though the minerals can be completely retained if the tubers are vacuum-packed (Elkins, 1979). The iron content of canned sweet potato increased threefold after 18 months of storage and was obviously derived from the metal can. The leaching loss on boiling potatoes could be minimized if the skin were retained, as reported by True *et al.* (1979) who found over 90 percent retention when potato was boiled for 14 minutes with the skin on. There are no leaching losses in the case of copper and zinc and so they present no problem (Finglas and Faulks, 1985).

In some traditional processing an appreciable amount of protein could be lost. For example in the preparation of *chuño blanco* the protein content of the potato is reduced from 2.1 percent to 1.9 percent (Table 6.3). Some of the loss is because of removal in the exudate, but most of it takes place during the soaking in water, about 50 percent. Most of the vitamins are also lost in the process. There is a 90 percent loss in vitamin B_1, 75 percent in B_2 and less than 50 percent of the niacin is retained. *Papa seca* , vitamins the best. There is an increase in the iron, calcium and phosphorus content in all the preparations (Table 6.3) because of the increased concentration of the product.

In the preparation of *gari* (Table 6.5) over one-third of the protein is lost, with greater losses for *fufu* and *lafun* (Oke, 1968). Most of the minerals are also reduced appreciably, except iron, which is increased, probably owing to the use of an iron pot for frying the product (Table 6.6). When yam is boiled, steamed or baked, the dietary fibre content rises because starch is modified and some minerals are lost, particularly phosphorus and potassium. (see Table 6.4.) Processing makes some difference to the percentage of nutrients that sweet potato will provide, as shown in Table 6.8. The 6.6 percent increase in the maltose content of sweet potato on cooking is not typical for other root crops, which presumably contain less amylases (Tamate and Bradbury, 1985).

TABLE 6.3

Composition of raw potato, *chuño* and *papa seca* (per 100g)

Product	Energy (kJ)	Energy (Kcal)	Crude protein (g)	Carbohydrate (g)	Ca (mg)	P (mg)	Fe (mg)	Thiamin (mg)	Riboflavin (mg)	Niacin (mg)	Ascorbic acid (mg)
Raw potato	335	80	2.1	18.5	9	50	0.8	0.10	0.04	1.50	20
Chuño blanco	1 351	323	1.9	77.5	92	54	3.3	0.03	0.04	0.38	1.1
Chuño negro	1 393	323	4.0	79.4	44	203	0.9	0.13	0.17	3.40	1.7
Papa seca	1 347	322	8.2	72.6	47	200	4.5	0.19	0.09	5.00	3.2

Source: Woolfe, 1987.

TABLE 6.4

Effect of cooking on composition of yam

(results calculated on fresh weight basis[1])

	Analysis of control (g/kg^{-1})	Differences		
		Boiled-control	Steamed-control	Baked-control
Moisture	766 (12)[2]	12.0*	-1.8	-68.0**
Ash	7.5 (0.3)	-1.2**	-0.1	0.1
Starch	186 (21)	5.8	-3.1	-3.6
Dietary fibre	15.6 (4.4)	16.3**	16.0**	9.2*
Sugars				
Fructose	2.2 (0.9)	-0.7	-0.6	-0.8
Glucose	1.6 (0.9)	-0.4	-0.5	-0.6
Sucrose	5.1 (2.4)	1.4	0.7	0.9
Maltose	0.8 (0.3)	0.1	-0.2	-0.2*
Minerals mg/kg^{-1}				
Ca	60 (12)	-2.6	-9.9*	-4.7
P	390 (20)	-33.0**	8.4	-25
Mg	150 (10)	-8.0	2.2	-11.4
Na	58 (25)	-28.0*	-17*	-8
K	3 450 (200)	-630.0**	-70	-230
S	140 (10)	-17.0**	2.4	-1.0
Zn	3.2 (0.3)	0.1	-0.1	-0.3*
Mn	0.3 (0.1)	-0.1	-0.1	-0.1
Al	2.1 (1.1)	0	0.2	0.3
B	1.0 (0.1)	-0.2*	-0.1	-0.1

[1]Results of five tubers of cultivar Da10 were averaged, standard deviations given in parentheses, differences marked with one asterisk are significant at $P<0.05$, those marked with two asterisks are significant at $P<0.01$. Other results not included above are protein 17.8 (3.9), fat 0.6 (0.5), raffinose 0.4 (0.3) g/kg^{-1}; Fe 6.5 (3.9), Cu 1.7 (0.3) mg/kg^{-1}.

[2]The value of 766 was the moisture content at harvest in Western Samoa; the moisture content before cooking in Canberra was 752 (16) g/kg^{-1}.

Source: Bradbury & Holloway, 1988.

TABLE 6.5

Proximate analysis of cassava and its products

(percentage of dry matter)

	Dry matter	Crude protein	Ether extract	Crude fibre	Carbo-hydrate	Ash	Calories
Cassava	28.5	2.6	0.46	0.43	94.1	2.4	391
Gari	85.6	0.9	0.10	0.40	81.8	1.4	323
Fufu	4.7	0.6	0.14	0.20	95.8	0.5	393
Lafun	80.5	0.8	0.40	0.73	96.4	2.0	391
Kpokpogari	87.8	1.5	0.0	4.2	78.1	5.2	312

Source: Oke, 1968.

TABLE 6.6

Minor elements in cassava and its products in Nigeria

Food stuff	Fraction in p.p.m. of dry matter								Dry matter (percentage)			
	Na	Mn	Fe	Cu	B	Zn	Mo	Al	P	K	Ca	Mg
Cassava	56	12	18	8.4	3.3	24	0.9	19	0.15	1.38	0.13	0.04
Gari	74	12	22	4.3	6.6	19	0.7	30	0.04	0.52	0.07	0.00
Fufu	36	8	62	3.0	8.5	11	0.9	15	-	-	-	-
Lafun	54	12	66	5.0	9.5	19	1.0	125	-	-	-	-
Kpokpogari	74	1.0	12	3.0	3.3	19	1.0	165	-	-	-	-
Yam	22	8	8	8	9	17	0.9	15	0.09	1.5	0.16	0.05

Source: Oke, 1968

TABLE 6.7

Effect of cooking on composition of sweet potato

(fresh weight basis)

	Analysis of control[1] g/kg-1	Difference[2]		
		Boiled-control	Steamed-control	Baked-control
Moisture	684 (29)[3]	43.0**	16.0**	-73.0**
Ash	7.6 (0.7)	-1.2**	-0.7*	0.4
Starch	213 (18)	-98.0**	-62.0**	-119.0**
Dietary fibre	14 (2.0)	20.6**	20.7**	11.2*
Sugars[4]				
Fructose	3.3 (1.2)	-0.8*	-0.4	-0.7*
Glucose	4.5 (1.1)	-0.6	-0.4	-0.8
Sucrose	20.3 (5.8)	1.1	1.9	4.0
Maltose	6.4 (10.2)	64.3**	68.8**	64.5**
Minerals mg/kg^{-1}				
Ca	450 (60)	5	-67	-20
P	290 (30)	10	14	10.0*
Mg	360 (60)	28	-37	-6
Na	730 (160)	-127	-104	-27
K	2 430 (190)	-360	470.0*	370
S	130 (20)	11	11	8
Zn	2.9 (0.7)	-0.5**	0.1	0.6
Mn	2.6 (1.4)	0.1	-0.3	-0.1
Al	2.4 (1.2)	1.8	-1.0	-0.3
B	1.4 (0.2)	0.0	-0.1	-0.1

[1]Other results are crude protein 17.7 (2.4) g/kg^{-1}, Fe 7.0 (2.6), Ca 2.2 (0.6) mg/kg^{-1}. Standard deviations given in parentheses.
[2]Results of five tubers were averaged from three tubers of 83003-15, one tuber of each of 83003-13 and Hawaii. Differences showing one asterisk indicate a significant ($P<0.05$) change on cooking, two asterisks indicate $P<0.01$.
[3]The value of 684 for moisture was that on harvesting in Tonga: the moisture content before cooking in Canberra was 634 (30) g/kg^{-1}.
[4]Total sugar: control 34.5, boiled 98.5, steamed 104.4, baked 101.5 g/kg^{-1}.
Source: Bradbury and Holloway, 1988.

TABLE 6.8

Percentages of adult recommended daily allowances provided by 100 g servings of processed potato products[1]

Potato product	Crude protein	Thiamin	Niacin	Folic acid	Pyridoxine[2]	Ascorbic acid	Iron
Boiled in skin[3]	6	8	8	7	11	50	7-12
Frozen, mashed reheated	5	5	4	-	-	13	7-12
French fries, finish-fried	8	8	11	6	18	40	11-20
Chips[4]	5	6	8	3	13	19	8-14
Flakes (prepared)	5	0-3	5	-	-	17	3-6
Granules (prepared)	5	0-3	4	3	8	10	6-10
Canned (solids)	3	3	4	6	7	40	3-6

[1]Unless otherwise indicated, calculated from figures for processed potato products given in Table 6.1 as percentages of RDAs given by Passmore *et al.* (1974).
[2]As percentage of USA recommended daily allowance.
[3]Domestic preparation.
[4]A 33.3 g serving, considered to be a more realistic estimate of a single serving of chips.
Source: Woolfe, 1987.

7. Toxic substances and antinutritional factors

Root crops, in common with most plants, contain small amounts of potential toxins and antinutritional factors such as trypsin inhibitors. Apart from cassava, which contains cyanogenic glucosides, cultivated varieties of most edible tubers and roots do not contain any serious toxins. Wild species may contain lethal levels of toxic principles and must be correctly processed before consumption. These wild species are useful reserves in times of famine or food scarcity. Local people are aware of the potential risks in their use and have developed suitable techniques for detoxifying the roots before consumption.

CASSAVA TOXICITY

The main toxic principle which occurs in varying amounts in all parts of the cassava plant is a chemical compound called linamarin (Nartey, 1981). It often coexists with its methyl homologue called methyl-linamarin or lotaustralin. Linamarin is a cyanogenic glycoside which is converted to toxic hydrocyanic acid or prussic acid when it comes into contact with linamarase, an enzyme that is released when the cells of cassava roots are ruptured. Otherwise linamarin is a rather stable compound which is not changed by boiling the cassava. If it is absorbed from the gut to the blood as the intact glycoside it is probably excreted unchanged in the urine without causing any harm to the organism (Philbrick, 1977). However, ingested linamarin can liberate cyanide in the gut during digestion.

Hydrocyanic acid or HCN is a volatile compound. It evaporates rapidly in the air at temperatures over 28°C and dissolves readily in water. It may easily be lost during transport, storage and analysis of specimens. The normal range of cyanogen content of cassava tubers falls between 15 and

400 mg HCN/kg fresh weight (Coursey, 1973). The concentration varies greatly between varieties (Fig. 7.1) and also with environmental and cultural conditions. The concentration of the cyanogenic glycosides increases from the centre of the tuber outwards (Bruijn, 1973). Generally the cyanide content is substantially higher in the cassava peel. Bitterness is not necessarily a reliable indicator of cyanide content.

Traditional processing and cooking methods for cassava can, if efficiently carried out, reduce the cyanide content to non-toxic levels. An efficient processing method will release the enzyme linamarase by disintegrating the

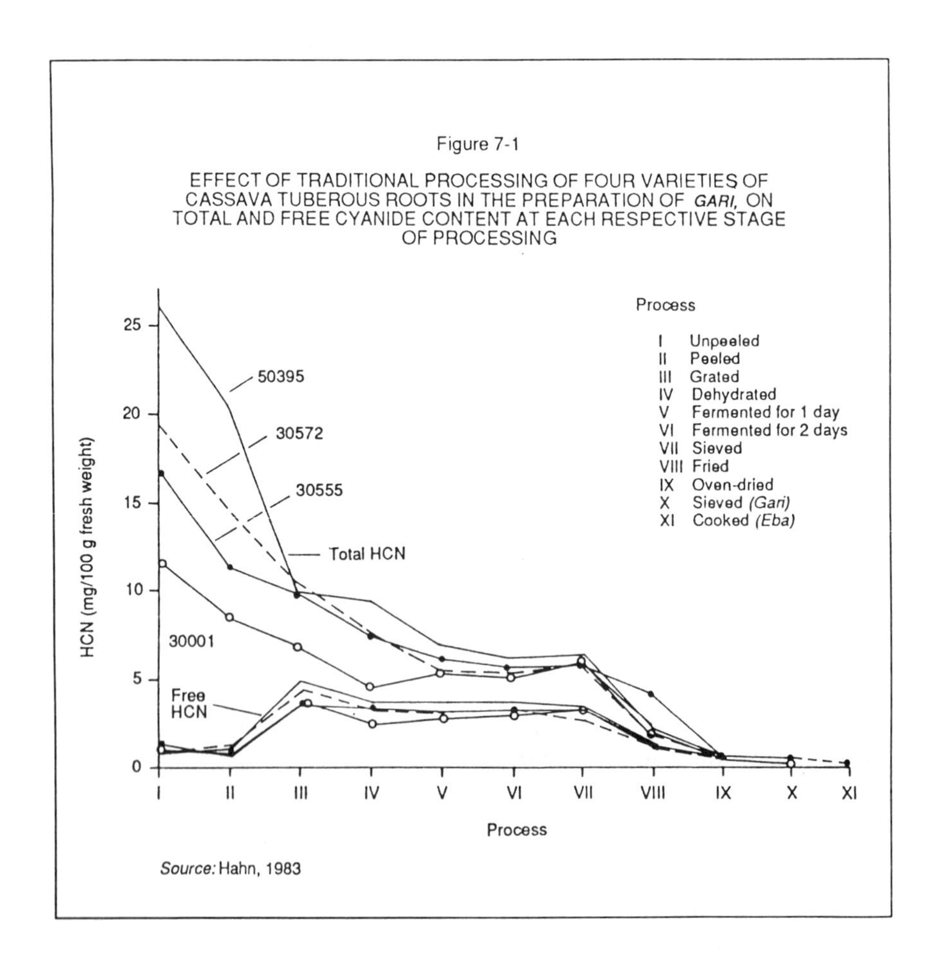

Figure 7-1

EFFECT OF TRADITIONAL PROCESSING OF FOUR VARIETIES OF CASSAVA TUBEROUS ROOTS IN THE PREPARATION OF *GARI*, ON TOTAL AND FREE CYANIDE CONTENT AT EACH RESPECTIVE STAGE OF PROCESSING

Source: Hahn, 1983

microstructure of the cassava root. On bringing this enzyme into contact with linamarin the glucoside is converted into hydrogen cyanide. The liberated cyanide will dissolve in the water when fermentation is effected by prolonged soaking, and will evaporate when the fermented cassava is dried. Sun drying fresh cassava pieces for short periods is an inefficient detoxification process. Cyanide will not be completely liberated and the enzyme will be destroyed during drying. Sun drying processing techniques reduce only 60 to 70 percent of the total cyanide content in the first two months of preservation. Cyanide residues can be quite high in the dry tubers, from 30 to 100 mg/kg (Casadei, 1988). Simple boiling of fresh root pieces is not always reliable since the cyanide may be only partially liberated, and only part of the linamarin may be extracted in the cooking water. The reduction of cyanides depends on whether the product is placed in cold water (27°C) or directly into boiling water (100°C). After 30 minutes cooking, the remaining cyanides are, in the first case, 8 percent of the initial value, and in the second case about 30 percent (Essers, 1986).

Various authors have suggested different minimal levels for toxicity. Rosling (1987) was of the opinion that an intake of over 20 mg per 100 g of cassava is toxic, while Bolhuis (1954) set the toxic level at an intake of 50 to 60 mg daily for a European adult.

Table 7.1 shows the HCN content of various processed cassava products. It indicates that a dramatic reduction in the hydrocyanic acid content of the raw cassava has occurred during processing. Soaking in water improves detoxification as cells are broken by osmosis and fermentation, which facilitates hydrolysis of the glycosides. Short soaking (four hours) is ineffective, but when longer periods are used (18 to 24 hours) cyanide levels can be reduced by 50 percent (Table 7.2). Squeezing the product is a fundamental step in the elimination of the soluble cyanides.

Pathophysiology of cyanide intoxication

Cyanide is detoxicated in the body by conversion to thiocyanate, a sulphur-containing compound with goitrogenic properties. The conversion is catalysed by an enzyme thiosulphate cyanide sulphur transferase (rhodanase) present in most tissues in humans, and to a lesser extent by mercapto-

TABLE 7.1

HCN content of various cassava products during processing

Food item	Detoxification stage	Remaining HCN Mean (mg/kg)	(percentage)
Mpondu	Fresh leaves	68.6	100.0
	Washed leaves (cold water)	63.9	93.1
	Dried leaves	66.1	96.3
	Boiled leaves (15 min in water)	3.7	5.4
	Boiled leaves (30 min in water)	1.2	1.7
	Boiled cassava		
	Fresh roots (sweet)	10.7	100.0
	Boiled roots (20 min in water)	1.3	12.1
Fufu	Fresh roots (sweet and bitter)	111.5	100.0
	Soaked roots (3 days)	19.4	17.4
	Dried roots (3 days)	15.7	14.1
	Uncooked *fufu* (flour and water)	2.5	2.2
	Cooked *fufu*	1.5	1.3
Fuku	Fresh roots (sweet)	25.5	100.0
	Uncooked *fuku* (heated)	4.2	16.4
	Cooked *fuku*	1.2	4.7
Gari	Mash	90.1	100.0
	24 h fermentation	73.2	81.2
	48 h fermentation	55.3	61.3
	48 h pressing	36.0	40.0
	Roasting	25.8	28.6
Lafun	Mash	16.5	100.0
	5-day soaking	35.9	21.8
	5-day soaking + 48 h drying	25.5	15.5
	5-day soaking + 96 h drying	19.6	11.9

Source: Bourdoux *et al.* 1982; Oke, 1984.

pyruvate cyanide sulphur transferase which is present in red blood cells (Fielder and Wood, 1956). The essential substrates for conversion of cyanide to thiocyanate are thiosulphate and 3-mercaptopyruvate, derived mainly from cysteine, cystine and methionine, the sulphur-containing amino-acids. Vitamin B_{12} in the form of hydroxycobalamin probably influences the conversion of cyanide to thiocyanate. Hydroxycobalamin has been reported to increase the urinary excretion of thiocyanate in experimental animals given small doses of cyanide (Wokes and Picard, 1955; Smith and Duckett, 1965). About 60 to 100 percent of the injected cyanide in toxic concentration is converted to thiocyanate within 20 hours and enzymatic conversion accounts for more than 80 percent of cyanide detoxification (Wood and Cooley, 1956). Thiocyanate is widely distributed throughout body fluids including saliva, in which it can readily be detected. In normal health, a dynamic equilibrium between cyanide and thiocyanate is maintained. A low protein diet, particularly one which is deficient in sulphur-containing amino-acids may decrease the detoxification capacity and thus make a person more vulnerable to the toxic effect of cyanide (Oke 1969, 1973). Excessive consumption of cassava, as the sole source of dietary energy and main source of protein, could thus increase vulnerability to cyanide toxicity.

Diseases related to cassava toxicity

Several diseases have been associated with the toxic effects of cassava. Its causative role has been confirmed in the pathological condition of acute cyanide intoxication and in goitre. There is also some evidence linking two types of paralysis to the combined effects of a high cyanide and low sulphur intake, such as could result from a diet dominated by inefficiently processed cassava. In these two diseases, tropical atoxic neuropathy and epidemic spastic paraparesis, paralysis follows damage to the spinal cord. The role of cyanide toxicity in the causation of tropical diabetes, and in congenital malformation has not been established. Similarly its supposed beneficial effects on sickle cell anaemia, shistosomiasis and malignancies are still hypothetical.

Acute cyanide intoxication. Symptoms appear four to six hours after a meal of raw or insufficiently processed cassava and consist of vertigo, vomiting, collapse and in some cases death within one or two hours. Treatment is quite effective and cheap. The principle is to increase the detoxicating capacity of the patient by giving an intravenous injection of thiosulphate and thereby making more sulphur available for conversion of cyanide to thiocyanate.

Endemic goitre. Cyanide taken in the diet is detoxified in the body, resulting in the production of thiocyanate. Thiocyanate has the same molecular size as iodine and interferes with iodine uptake by the thyroid gland (Bourdoux *et al.*, 1978). Under conditions of high ingestion of inefficiently processed cassava, there may be a chronic cyanide overload leading to a high level of serum thiocyanate of 1 to 3 mg/100 ml, compared to a normal level of about 0.2 mg/100 ml. Under such conditions there is an increased excretion of iodine and a reduced iodine uptake by the thyroid gland, resulting in a low thiocyanate/iodine (SCN/I) excretion ratio. The value of the threshold level for this ratio seems to be three (Delange *et al.*, 1983) after which endemic goitre appears. This phenomenon can occur only when the iodine intake is below about 100 mg per day. At SCN/I ratios of lower than two there is a risk of endemic cretinism, a condition characterized by severe mental retardation and severe neurologic abnormalities (Ermans *et al.*, 1983).

Studies in Zaire have shown that the population of Ubangi, who consume a high amount of sun dried but unfermented cassava products, have a low SCN/I ratio of 2 to 4 and suffer from endemic goitre and cretinism. Whereas in Kirn, where fermented and dried cassava paste is eaten, the SCN/I ratio goes up to three to five and there is a low incidence of goitre. In Bas Zaire, where properly processed cassava products are eaten, the SCN/I ratio is higher than seven and there is no goitre. A low ratio leads to abnormal levels of the thyroid stimulating hormone (TSH) and low thyroxine (T4). Ayangade *et al.*, (1982) found that in pregnant women the thiocyanate level of the cord blood was proportional to the maternal serum thiocyanate level, indicating that thiocyanate can cross the placental barrier and affect the foetus. However, there is very little thiocyanate in breast milk indicating that the mammary gland does not concentrate thiocyanate and so breast-fed infants are not affected.

When iodine supplements are given, for example, by adding potassium iodide to local supplies of salt, goitre is reduced in spite of a continued high intake of cassava products. Where salt intake is small or variable, iodized oil, given by mouth, provides protection for one to two years. In the Amazon jungle some tribal people eat as much as one kg of cooked fresh cassava per person per day and consume up to three litres of fermented cassava beer, but there have been no reported cases of either goitre or ataxic neuropathy. These tribes also consume a considerable amount of animal and fish protein and thus have high levels of sulphur-amino acids and iodine in their diet.

Neurological disorders

Cyanide intake from a cassava-dominated diet has been proposed as a contributing factor in two forms of nutritional neuropathies, tropical ataxic neuropathy in Nigeria (Osuntokun, 1981) and epidemic spastic paraparesis (Cliff *et al.*, 1984). These disorders are also found in some cassava growing-areas of Tanzania and Zaire.

Tropical ataxic neuropathy. This disease is common in a particular area in Nigeria where a lot of cassava is consumed without the addition of sufficient protein-rich supplementary foods to provide an adequate supply of sulphur amino-acids for the detoxification of ingested cyanide. The consumed cassava product, called *purupuru,* is processed by an insufficient fermentation of the cassava, which leaves a residual cyanide content of up to 0.10 M mole/g. As much as two kg of this foodstuff is consumed daily, leading to the ingestion of about 50 mg of cyanide. The toxic level for an adult is about 60 mg. The clinical picture is dominated by damage to one of the sensory tracts in the spinal cord resulting in an uncoordinated gait called ataxia.

When patients are brought to the hospital they have a high plasma thiocyanate level. On admission they are put on a hospital diet which is highly nutritious and includes cassava only twice a week. Within a short period the plasma thiocyanate level returns to normal, and the patients recover. However, on discharge, they go back to their original diet of cassava and so the condition reappears (Osuntokun, 1968).

All the cases reported came from the area where cassava is cultivated and eaten in large quantities, with no cases in the nearby areas where yam predominates. A change in the diet of the population at risk in Nigeria has reduced the incidence of this disease.

Epidemic spastic paraparesis. This is a situation of depending on very toxic varieties of cassava as a food security crop (Cliff *et al.*, 1984). In parts of Mozambique a bitter toxic type of cassava is often planted as a food reserve because of its high yield. As cassava constitutes about 80 percent of the basic diet, there is normally a standard method of preparation which makes the cassava safe for consumption. Cassava, containing about 327 mg HCN/kg, is peeled, sliced and sun dried for about three weeks after which the cyanide level is reduced to about 95 mg/kg. It is then pounded to a flour which is mixed with hot water to make a paste called *chima*. This paste is normally eaten with a relish of beans, fish or vegetables, to provide a well-balanced meal.

During a prolonged period of drought all the food crops in this area were lost except the toxic variety of cassava. The foodstores were depleted and many families had no alternative, but to resort to the toxic cassava. Normal processing time was reduced because of the emergency and so there was no proper detoxification. The people knew this but they had no other choice of action except to die of starvation. On eating the underprocessed *chima* without their usual protein-rich supplement they complained that it was more bitter than normal. After about four to six hours they suffered from nausea, vertigo and confusion. Sufferers showed a high serum thiocyanate level and a urinary thiocyanate excretion of about ten times that of non-cassava-eating groups in Mozambique. There followed a sudden appearance of many cases of spastic paraparesis, indicating an extensive epidemic. This disease affects mainly women and children. It damages the nerve tract in the spinal cord that transmits signals for movement, thus causing a spastic paralysis of both legs (Rosling, 1983). Outbreaks have been reported during the dry season from two areas in Zaire (Nkamany and Kayinge, 1982) and during droughts in one area in Mozambique (Cliff *et al.*, 1984) and one area in Tanzania (Howlett, 1985).

During these drought periods about 500 g of dried cassava, or 1.5 kg on a fresh weight basis, is consumed daily, representing an intake of 1 500 kcal and 50 mg cyanide per day. This level approaches the toxic level of 60 mg. The body can safely detoxify about 20 mg cyanide per day but when this level increases to 30 mg symptoms of acute intoxication develop in many consumers and hence the epidemics. If there is a period during which a high cassava intake and a low protein-rich food intake, to supply sulphur amino-acids for detoxification, coincide, this combination precipitates the outbreak of this disease. The situation may be compared to the epidemics of lathyrism that occured in drought-affected areas of India owing to the high-level intake of the drought-resistant pea, *Lathyrus sativa.*

Production of low-cyanide foods

The development of a more sensitive method for cyanide determination in foods by Cooke (1978a) and an in-depth study of some traditional cassava foods have led to a better understanding of the detoxification mechanism of cyanide in foods and to improved recommendations for processing cassava.

Cyanide occurs in cassava and cassava products in two forms, the glucosidic form, which is the linamarin itself, and the non-glucosidic or bound form which is cyanohydrin. Under normal conditions of hydrolysis, when the enzyme linamarase reacts with linamarin, it is hydrolysed to cyanohydrin which, on decomposition, gives acetone and hydrocyanic acid. However, under acid conditions, of pH4 or less, which tend to occur in some lactic acid fermentations of cassava, the cyanohydrin decomposition is hindered and it becomes stable. It is relatively easy to get rid of free cyanide, which is present at about 10 percent in both peeled and fresh cassava, especially in solution, but the non-glucosidic cyanide may hydrolyse very slowly and result in a lot of residual cyanide in cassava products. Thus drying cassava chips in an air oven at 47° and 60°C causes a decrease in the bound cyanide content of 25 to 30 percent, whereas faster drying at 80°C or 100°C gave only a 10 to 15 percent decrease of the bound cyanide. However, losses of free cyanide were 80 to 85 percent and 95 percent respectively (Cooke and Maduagwu, 1978b). Drying results in an apparent increase in cyanide concentration because of loss of water (Bourdoux *et al.*, 1982). The longer the drying the higher the amount of water removed. About 14 percent of the water can be removed during the first day, reaching a level of up to 70 percent after eight days. This leads to an increase in cyanide concentration from 70 mg/kg on the first day to 91 mg/kg after eight days.

Soaking in water at 30°C, boiling or cooking removes free cyanide but only about 55 percent of the bound cyanide is released after 25 minutes. However, the bound cyanide is removed by prolonged soaking as fermentation begins (Table 7.2) through the action of the enzyme linamarase which is released by disruption of the tuberous tissues. If water is added at this stage most of the cyanide is removed. Meuser and Smolnik (1980) were able to improve the production of *gari* by washing the mash after fermentation to remove the residual bound cyanide which was still present as cyanohydrin because of its higher stability at the lower pH.

The result of different drying techniques is shown in Table 7.3. Freeze-drying or flash-drying eliminated only the free cyanide, which accounted for about 50 percent of the total cyanide present. Roller-drying of the fresh pulp at a pH of 5.5 to 5.7 removed virtually all the cyanide, whereas if the fermented pulp was dried on rollers or on drums high amounts of cyanide were retained in the dried product because of the acid condition (pH 3.8) of the fermented pulp. In the detoxification of cassava products fermentation is most effective when accompanied by squeezing and washing of the acidic pulp. Residual cyanide can be reduced further by sun drying or frying. This had been confirmed by Hahn (1983) as shown in Fig. 7.1. In traditional preparations of various food products from cassava, there may be some residual cyanide because of insufficient tissue disintegration during processing and insufficient washing. It is the residual cyanide that is responsible for toxicity. Some of these preparations have been simulated in the laboratory and modified to give much lower cyanide levels (Bourdoux *et al.*, 1983).

TABLE 7.2
Effects of soaking on the HCN content of six bitter cassava roots

Soaking period (days)	Remaining HCN (percentage)
0	100.0
1	55.0
2	42.3
3	19.0
4	10.9
5	2.7

Source: Bourdoux *et al.*, 1983

SWEET POTATO

Sweet potato contains raffinose, one of the sugars responsible for flatulence. Three of the sugars which occur in plant tissues, raffinose, stachyose and verbascose are not digested in the upper digestive tract, and so are fermented by colon bacteria to yield the flatus gases, hydrogen and carbon dioxide. The level of raffinose present depends on the cultivar. In some parts of Africa the cultivars used are considered too sweet and cause flatulence (Palmer, 1982), Lin *et al.* (1985) have established that sweet potato shows trypsin inhibitor activity (TIA) ranging from 90 percent inhibition in some varieties to 20 percent in others. There is a significant correlation between the trypsin inhibitor content and the protein content of the sweet potato variety. Heating to 90°C for several minutes inactivates trypsin inhibitors. Lawrence and Walker (1976) have implicated TIA in sweet potato as a contributory factor in the disease enteritis necroticans. This seems doubtful since sweet potato is not usually eaten raw and the activity of the trypsin inhibitor present is destroyed by heat.

In response to injury, or exposure to infectious agents, in reaction to physiological stimulation or on exposure of wounded tissue to fungal contamination, sweet potato will produce certain metabolites. Some of

TABLE 7.3

Effect of drying on HCN content of cassava

Drying process		HCN (ppm)
Freeze drying	Pulp	439
Flash drying	Slices	432
Air drying 40°C	Chips, pulp	13
Heated air drying 180°C	Chips	14
	Fermented pulp	77
Drum drying	Pulp	8
	Fermented pulp	121
HCN of pulp	free and bound	900

Source: Meuser & Smolnik, 1980.

these compounds, especially the furano-terpenoids are known to be toxic (Uritani, 1967). Fungal contamination of sweet potato tubers by *Ceratocystis fimbriata* and several *Fusarium* species leads to the production of ipomeamarone, a hepatoxin, while other metabolites like 4-ipomeanol are pulmonary toxins. Baking destroys only 40 percent of these toxins. Catalano *et al.* (1977) reported that peeling blemished or diseased sweet potatoes from 3 to 10 mm beyond the infested area is sufficient to remove most of the toxin.

POTATO

Potato contains the glycoalkaloids alpha-solanine and alpha-chaconine (Maga, 1980), concentrated mainly in the flowers and sprouts (200 to 500 mg/100 g). In healthy potato tubers the concentration of the glycoalkaloids is usually less than 10 mg/100 g and this can normally be reduced by peeling (Wood and Young, 1974; Bushway *et al.*, 1983). In bitter varieties the alkaloid concentration can go up to 80 mg/100 g in the tuber as a whole and up to 150-220 mg/100 g in the peel. The presence of these glycoalkaloids is not perceptible to the taste buds until they reach a concentration of 20 mg/

100 g when they taste bitter. At higher concentrations they cause a burning and persistent irritation similar to hot pepper. At these concentrations solanine and other potato glycoalkaloids are toxic. They are not destroyed during normal cooking because the decomposition temperature of solanine is about 243°C.

Levels of glycoalkaloids may build up in potatoes which are exposed to bright light for long periods. They may also result from wounding during harvest or during post-harvest handling and storage, especially at temperatures below 10°C (Jadhav and Salunkhe, 1975). Glycoalkaloids are inhibitors of choline esterase and cause haemorrhagic damage to the gastro-intestinal tract as well as to the retina (Ahmed, 1982). Solanine poisoning has been known to cause severe illness but it is rarely fatal (Jadhav and Salunkhe, 1975).

Potato also contains proteinase inhibitors which act as an effective defense against insects and micro-organisms but are no problem to humans because they are destroyed by heat. Lectins or haemogglutenins are also present in potato. These toxins are capable of agglutinating the erythocytes of several mammalian species including humans (Goldstein and Hayes, 1978), but this is of minimal nutritional significance as haemogglutenins are also destroyed by heat, and potatoes are normally cooked before they are eaten.

COCOYAM

The high content of calcium oxalate crystals, about 780 mg per 100 g in some species of cocoyam, *Colocasia* and *Xanthosoma*, has been implicated in the acridity or irritation caused by cocoyam. Oxalate also tends to precipitate calcium and makes it unavailable for use by the body. Oke (1967) has given an extensive review of the role of oxalate in nutrition including the possibility of oxalaurea and kidney stones. The acridity of high oxalate cultivars of cocoyam can be reduced by peeling, grating, soaking and fermenting during processing.

Acridity can also be caused by proteolytic enzymes as in snake venoms. Attempts have been made to isolate such enzymes from taro, *Colocasia esculenta*, and the principal component has been called "taroin" by Pena *et al.* (1984).

BANANA AND PLANTAIN

Banana and plantain do not contain significant levels of any toxic principles. They do contain high levels of serotonin, dopamine and other biogenic amines. Dopamine is responsible for the enzymic browning of sliced banana. Serotonin intake at high levels from plantain has been implicated in the aetiology of endomyocardial fibrosis (EMF) (Foy and Parratt, 1960). However, Ojo (1969) has shown that serotonine is rapidly removed from the circulating plasma and so does not contribute to elevated levels of biogenic amines in healthy Nigerians. It has been confirmed by Shaper (1967) that there is insufficient evidence for regarding its level in plantain as a factor in the aetiology of EMF.

YAM

The edible, mature, cultivated yam does not contain any toxic principles. However, bitter principles tend to accumulate in immature tuber tissues of *Dioscorea rotundata* and *D. cayenensis*. They may be polyphenols or tannin-like compounds (Coursey, 1983). Wild forms of *D. dumetorum* do contain bitter principles, and hence are referred to as bitter yam. Bitter yams are not normally eaten except at times of food scarcity. They are usually detoxified by soaking in a vessel of salt water, in cold or hot fresh water or in a stream. The bitter principle has been identified as the alkaloid dihydrodioscorine, while that of the Malayan species, *D. hispida*, is dioscorine (Bevan and Hirst, 1958). These are water soluble alkaloids which, on ingestion, produce severe and distressing symptoms (Coursey, 1967). Severe cases of alkaloid intoxication may prove fatal. There is no report of alkaloids in cultivated varieties of *D. dumetorum*.

Dioscorea bulbifera is called the aerial or potato yam and is believed to have originated in an Indo-Malayan centre. In Asia detoxification methods, involving water extraction, fermentation and roasting of the grated tuber are used for bitter cultivars of this yam. The bitter principles of *D. bulbifera* include a 3-furanoside norditerpene called diosbulbin. These substances are toxic, causing paralysis. Extracts are sometimes used in fishing to immobolize the fish and thus facilitate capture. Toxicity may also be due to saponins in the extract. Zulus use this yam as bait for monkeys and hunters in Malaysia use it to poison tigers. In Indonesia an extract of *D. bulbifera* is used in the preparation of arrow poison (Coursey, 1967).

PHYTATE

Phytate is a storage form of phosphorus which is found in plant seeds and in many roots and tubers (Dipak and Mukherjee, 1986). Phytic acid has the potential to bind calcium, zinc, iron and other minerals, thereby reducing their availability in the body (Davis and Olpin, 1979; O'Dell and Savage, 1960). In addition, complex formation of phytic acid with proteins may inhibit the enzymatic digestion of the protein (Singh and Krikorian, 1982). Iron and zinc deficiencies occur in populations that subsist on unleavened whole grain bread and rely on it as a primary source of these minerals. Deficiencies have been attributed to the presence of phytates.

Recently Marfo and Oke (1988) have shown that cassava, cocoyam and yam contain 624 mg, 855 mg and 637 mg of phytate per 100 g respectively (Table 7.4). Fermentation reduced the phytate level by 88 percent, 98 percent and 68 percent respectively, reduction being rapid within 48 hours but very slow after 72 hours processing. Thus processing into fermented foods will reduce the phytate level of root crops sufficiently to nullify its adverse effect. The loss of phytate during fermentation is due to the enzyme phytase, naturally present in the tubers or secreted by fermentative micro-organisms. Processing into *nbo* or *kokonte* resulted in a loss of only 18 percent of phytate in cassava and 30 percent each in cocoyam and yam (Table 7.5). Oven-drying has only a small reductive effect on the phytate content compared with fermentation. Cooking also has a significant effect, resulting in decrease of phytate of 62 percent, 65 percent and 68 percent respectively in yam, cocoyam and cassava.

TABLE 7.4
Phytate content of some unfermented and fermented tubers (mg/g)

Sample	Unfermented meal	Fermented meals (24 h)	(48 h)	(72 h)	(96 h)	Percentage loss[1]
Cassava	624	116	99	90	70	88.7
Cocoyam	855	180	28	13	13	98.4
Yam	637	394	296	222	211	66.8

[1]Percentage loss in phytate is the decrease in phytate after 96 hours fermentation expressed as a percentage of total phytate.
Source: Marfo & Oke, 1988.

TABLE 7.5
Effect of processing on phytate in cassava, cocoyam and yam

	Fresh and unprocessed	Sliced and cooked *(Ampesi)*	Flour cooked into a paste *(Tuo, kokonte)*	Dried granular powder *(Gari)*	*Gari* made into a paste *(Eba)*	*Fufu* (cooked and pounded)
CASSAVA	624	196	411	70	55	188
percentage loss[1]	-	68.5	18.1	86.0	89.0	69.8
COCOYAM	855	302	592	9	8	281
percentage loss[1]	-	64.6	30.7	98.9	99.0	67.1
YAM	637	239	412	188	179	209
percentage loss[1]	-	62.4	30.8	70.4	71.8	67.1

[1]Percentage loss in phytate is the decrease in phytate resulting from each processing method expressed as a percentage of total phytate content.
Source: Marfo & Oke, 1988.

8. New frontiers for tropical root crops and tubers

In this chapter agro-industrial possibilities for the expanded utilization of root crops are reviewed. The success of efforts made to increase the production of tropical root crops and to promote their use as food will depend on market demand. Farmers will not be encouraged to produce a marketable excess if this leads to glut, spoilage, and low prices. Policy-makers should not only promote policies to increase consumption of root crops as human foods and as animal feeds, but should also support research that will extend the utilization of these root crops. Efforts should be made to promote new technologies, appropriate for use by the rural population, to produce a variety of processed foods from root crops. This strategy will generate employment and improve incomes in rural areas. If demand is stimulated farmers will be encouraged to produce more root crops which can be converted to animal feed and industrial uses. Demand can be stimulated by development in three main areas:

- commercial dehydration of the root crops to produce flakes and flour that can be made into other food products;
- use of root crops as sources of industrial raw materials; and
- use of root crops as animal feed.

COMMERCIAL DEHYDRATION OF ROOT CROPS AND THEIR USE

The peeled root is rinsed to remove excess starch, then cut into slices, blanched, blended to a puree and dried. Peeling can be effected by immersion in 10 percent lye solution or by steaming at high temperatures (150°C) for short periods.

The drier, which may be a heat exchanger or drum drier, can be fired by agricultural wastes such as coconut husks which are abundant and cheap in

Southeast Asia. This reduces the moisture content from 70 percent to 12 percent. Improved preservation of root crops would increase their availability and reduce post-harvest wastage. Dried products require less storage space and have a longer shelf-life. They can be quickly reconstituted and prepared for eating, a factor of particular importance to urban consumers. Composite flour incorporating cocoyam has been used in extruded products such as noodles and macaroni. Similar processes could be used in the production of flour products from other root crops.

Processing will greatly increase the utilization of root crops. The flour can be used as a component of multimix baby foods and in composite flour for making bread. Research and development work on composite flour using root crops and other local products has advanced considerably in Colombia. Based on initial research in 1971-72, it was concluded that while rice and maize flours are preferable for use as non-wheat components in composite flours, cassava flour and starch also have good technical possibilities. The pilot work demonstrated that the production of bread from wheat flour diluted up to 30 percent with non-wheat components is possible on a commercial scale. But the large-scale introduction of such flours requires a concerted effort by both the public and private sectors to ensure the wide-spread availability at attractive prices of non-wheat raw materials. Expanded cassava production and lower prices are required if composite flour is to be economically attractive to millers and consumers (Goering, 1979).

Bread baked with composite flour from local resources would reduce the foreign exchange cost of imported wheat. This cost is particularly high inthe Philippines, where a processing plant has been set up to convert 5 000 kg of fresh sweet potato into flour every day. The bread from this flour will contain more calories and a higher content of vitamin A and lysine than wheat bread and will conserve foreign exchange. If it can be marketed at a reduced cost it may help to improve the nutritional status of the population. Taylor (1982) estimated that cash benefits to farmers producing raw material for this plant could substantially improve with a guaranteed market. Since the market and the financing of the crop is guaranteed, the promotion of sweet potato as a cash crop will be more easily accepted.

Fresh root crops are rarely exported in appreciable quantities because of their high water content and perishability. Cocoyam is exported in small amounts from Fiji, Western Samoa, Tonga and Cooke Island to the United States of America, New Zealand and Australia for the Polynesian and Melanesian immigrants. Yams are also exported from Latin America and Africa for immigrants in Europe, but quantities are small and prices are high.

USE OF ROOT CROPS AS A SOURCE OF INDUSTRIAL RAW MATERIAL

Most of the world's starch supplies are derived from either grains (corn, sorghum, wheat, rice), the major root crops (potato, sweet potato, cassava, arrowroot) or the pith of the sago palm. While starches from these various plant sources vary slightly in their physical and chemical properties, they can be substituted for each other across a wide spectrum of end uses. Cassava starch must compete with other starches and relative prices, quality and dependability of supplies are basic considerations in the determination of market shares (Goering, 1979).

Cassava tubers can be processed as a source of commercial starch for use in the foodstuff, textile and paper industries. As a foodstuff the starch may be converted by acid and enzyme hydrolysis to dextrins and glucose syrups, but maize starch is often available at a lower price for these purposes. The bland flavour of cassava starch, its low amylose content, non-retrogration tendency and excellent freeze-thaw stability makes it suitable for use in food processing. Simple modification of cassava starch by cross-bonding, or use of maize starch/cassava blends give properties ideally suited for use in a wide range of convenience foods. Starch from cocoyam has been recommended as a diluent in chemical and drug manufacturing and as a carrier in cosmetics such as face powder. It has a grain size similar to rice starch which is currently used for these purposes.

Cassava starch is manufactured in Thailand, Brazil and Malaysia and exported mainly to Japan and the United States of America. In 1975 the export level had reached about 100 000 tonnes per annum, equivalent in value to about US$30 million, with Thailand controlling about 50 percent of the market. In Thailand starch mills of various sizes have been set up including about 60 small mills with a unit capacity of two to three tonnes of starch per day, a similar number of modern mills producing 30 to 60 tonnes per day and a few industrial mills with a capacity of 100 to 150 tonnes per day. All these mills combine to give a total annual production of about 800 000 tonnes, of which about 700 000 tonnes are produced in modern mills. In Thailand a high proportion of the cassava starch produced can be utilized by local industries while the remainder is exported to other countries that have textile industries.

In several countries the traditional starch industry is an important source of starch for local users and provides a readily accessible market for tropical root crop production from small-scale farmers. A significant amount of rural non-farm employment also is generated by the industry. Factories typically are small scale (one tonne of raw root throughput per hour), are equipped with locally fabricated machinery and use crude sedimentation processes which result in a product of variable quality. This local industry frequently finds it difficult to compete with large-scale, semi-automated factories (throughput of up to 20 tonnes per hour) or with factories using grain, sometimes imported at low prices, as the raw material. Since the price of cassava roots is normally governed by the pellet industry, only limited possibilities exist to reduce starch prices in order to make exports more competitive. The excess capacity in the Thai starch industry places that country in a good position to meet the requirements of any future new markets elsewhere. The possible export of root starches is less attractive for other developing countries that do not have established positions in the market (Goering, 1979).

Starch can be hydrolysed to glucose and used as a sweetener. Starches from root crops are often more expensive than those obtained from cereals such as rice and maize. Increased production could possibly reduce root starch costs and make it more competitive. Fig. 8.1 presents a diagram of a potential agro-industrial system for cassava utilisation.

Yeast fermentation of the hydrolysed starch extract of cassava or other crops gives a good yield of absolute ethyl alcohol, which can be used as an extender of petroleum-based fuels by blending up to 20 percent, without

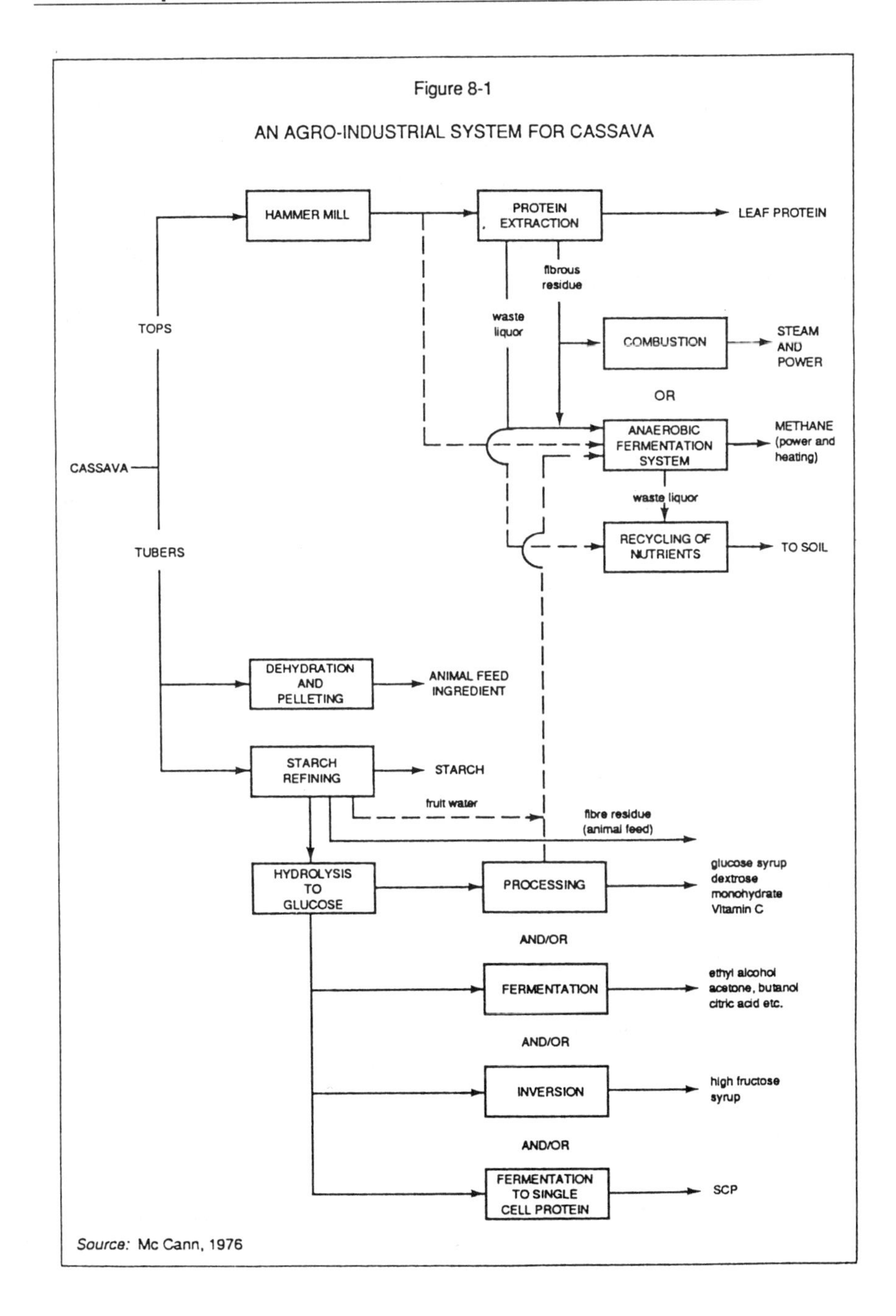
Figure 8-1
AN AGRO-INDUSTRIAL SYSTEM FOR CASSAVA
HAMMER MILL
PROTEIN EXTRACTION
LEAF PROTEIN
fibrous residue
waste liquor
TOPS
COMBUSTION
STEAM AND POWER
OR
ANAEROBIC FERMENTATION SYSTEM
METHANE (power and heating)
CASSAVA
waste liquor
RECYCLING OF NUTRIENTS
TO SOIL
TUBERS
DEHYDRATION AND PELLETING
ANIMAL FEED INGREDIENT
STARCH REFINING
STARCH
fruit water
fibre residue (animal feed)
HYDROLYSIS TO GLUCOSE
PROCESSING
glucose syrup
dextrose
monohydrate
Vitamin C
AND/OR
FERMENTATION
ethyl alcohol
acetone, butanol
citric acid etc.
AND/OR
INVERSION
high fructose syrup
AND/OR
FERMENTATION TO SINGLE CELL PROTEIN
SCP
Source: Mc Cann, 1976

altering the carburettor of most petrol engines. Brazil initiated a National Alcohol Programme in 1975 to produce ethyl alcohol from agricultural raw materials, much of it based on sugar cane. The technology is now well developed and production has started. The starch-to-alcohol conversion ratio is 1.76 kg starch for one litre of alcohol. Sugar cane is the most energy-efficient crop for alcohol production, but the use of cassava starch is increasing because it can be produced under conditions unsuitable for sugar cane. The cost from cassava was estimated at $0.57/gallon in 1978. PETROBRAS of Brazil was the first to test a large-scale plant to produce alcohol from cassava, with a full production capacity of 60 000 litres per day. Early runs were hampered by inadequate supplies of raw material and the high prices of cassava roots compared to the government-controlled price of petrol. During test runs 30 000 litres of alcohol were produced to specification. Doubling cassava yield would make the process more economical. This could lead to an increased production of cassava and its use as a renewable energy resource (Hammond, 1977).

Plans are under consideration in Indonesia to develop a number of commercial alcohol factories using raw material from sweet potato, cassava and sugar cane. Special attention will be given to sweet potato production because it can be harvested twice in a year compared to cassava which is harvested only once a year (Yang, 1982). Guaranteed markets are a great incentive to farmers to produce more root crops.

Fermentation of commercial starches with *Clostridium acetobutylicum* yields about 30 percent of the starch dry weight of mixed solvents of butanol, acetone and ethyl alcohol, from which the pure products can be obtained by distillation. A process based on high-yielding cassava cultivars as a starch source could be financially attractive.

Some root crops have considerable pharmaceutical potential not yet exploited in developing countries. Yam contains steroidal sapogenins which are useful starting materials for the preparation of cortisone and related drugs. In Mexico various wild species of *Dioscorea*, especially *D. mexicana,* contain useful quantities of these sapogenins as high percentage of the dry matter, and they can be converted into progesterone intermediates. Other species are good sources of diosgenin, a starting material for the manufacture of corticosteroids. Oral contraceptives based on progesterone are now widely used in family planning in several tropical countries. The possibility exists for their national production from local materials. This prospect has been extensively reviewed by Oke (1972).

Some species of *Dioscorea* grown in Southeast Asia contain toxic saponins. They are made locally into a medicinal shampoo which is used to destroy head lice. They are also used in an insecticidal powder, similar in effect to derris dust, which is used to destroy rice parasites in paddy fields in Malaysia. *D. cirrhosa* contains enough tannin for commercial use. Some cultivars of *D. alata* contain 6 to 38 percent of tannin which is used in Southeast Asia for tanning fishing nets and in Taiwan for tanning leather, to which it imparts a red colour (Coursey, 1967).

There are many traditional medicinal uses of some species of *Dioscorea* among Africans, Chinese and Asians which were discovered by trial and error. The Zulus use an extract of *D. sylvatica* for the treatment of uterine and mammary disorders in cattle. More work is needed in this area. Apart from the academic interest, the practical rewards in pharmacy and medicine could be far-reaching as in the case of the Indian snake-root, *Rauwolfia serpentina Beuth.* This was used for many years in traditional medicine by the Indians. It contains reserpine, which has found great use in modern medicine.

USE OF ROOT CROPS AS ANIMAL FEED

A factor restricting the development of animal production in many developing countries is the cost of imported feed which has often gone up several fold because of alterations in the rate of exchange of local currency with respect to world markets. As feed costs have increased, animal products have become very expensive. If part of the feed could be substituted with root crops such as cassava, then part of the maize ration could be freed for human consumption. Table 8.1 compares the nutritive value of different cassava products with sorghum and maize, as components of animal feeds. The low protein and fibre and high content of soluble carbohydrates (high digestibility) are notable features of the cassava root. Cassava tops, stems and leaves are also available as animal feed and are comparatively high in utilizable protein.

TABLE 8.1

Nutritive value of different cassava products compared with sorghum and maize (in percentages)

	Cassava chips	Cassava meal	Cassava refuse (fresh)	Cassava flour	Sorghum	Maize (ground)
Fresh						
moisture	11.7	11.2	80.0	14.9	11.9	13.4
C. protein	1.9	2.6	0.4	0.3	7.5	9.4
C. fibre	3.0	5.6	1.6	0.1	2.0	1.9
sol. carbohy.	80.5	73.9	17.6	84.4	74.6	70.1
ether ext.	0.72	0.55	0.10	0.10	2.32	3.64
ash	2.17	6.10	0.30	0.20	1.65	1.62
Dry						
total dry matter	88.3	88.6	20.0	85.1	88.1	86.6
C.I. (dry matter basis)	2.1	2.9	2.0	0.4	8.5	10.0
Calculated digestible nutrients						
C. protein	1.3	1.7	0.1	0.2	39.0	7.4
C. fibre	2.3	4.3	1.3	0.1	1.1	0.7
sol. carbohy.	78.9	72.4	9.9	86.3	48.5	64.5
ether ext.	0.36	0.28	0.10	0.10	1.35	2.18
Digestibility coefficient used						
C. protein	-	-	-	66	52	-
C. fibre	-	-	-	100	57	-
sol. carbohy.	-	-	-	99	65	-
ether ext.	-	-	-	100	58	-
Starch equivalent	83.2	78.7	11.5	84.1	89.5	78.2
Nutritive value	63.1	45.5	114.3	419.6	1.4	8.9

Source: H.K. Lim, 1967.

The International Development Research Centre in Canada sponsored a series of investigations into the use of cassava as animal feed. On the basis of their findings it is recommended that cassava could be a substitute of up to 40 percent for maize in the nutritionally balanced rations of pigs without any deleterious effect, and up to 30 percent in poultry rations.

Gómez *et al.* (1984) in Colombia reported that when cassava was substituted for corn in a poultry broiler ration at levels of up to 30 percent there was no significant difference in the performance at all levels, but the 20 percent level of substitution was the most economical. It required 215 kg of feed to produce 100 kg live weight with a 20 percent substitution, whereas it required 220 kg and 224 kg respectively for the corn feed and the 30 percent substitution feed as shown in Table 8.2. High levels of cassava intake are more acceptable for broiler production than for laying hens. Egg production and quality may be adversely affected by nutritional imbalances associated with rations high in cassava.

In the case of pigs (Table 8.3) the performance was progressively better as the level of cassava in the feed was increased. Thus it required 339 kg of feed to produce 100 kg weight with corn alone, whereas it required 337 kg and 331 kg respectively with 20 percent and 30 percent cassava substitution. In the economic assessment of the rations, the least-cost broiler diets containing 20 percent cassava meal gave the largest returns while profitability increased with the level of cassava meal in the case of the pig trials (Table 8.4), with those on the 30 percent cassava substitution being the most economical.

In view of the potential value of cassava to supply energy to dairy cattle, it has been used in a great number of experiments as the main source of energy, resulting in higher milk and fat yields and live weight gains (Pineda and Rubio, 1972). Similar results have been obtained for beef cattle when steers fed on commercial concentrate and cassava-based diets gained significantly faster than those fed bran or corn and cob-based diets. Better performance of bulls has also been reported by Montilla *et al.* (1975) on 40 percent cassava rations rather than on maize meal. Devendra (1977) has reported similar findings for goats and sheep, cassava enhancing utilization and hence nitrogen retention.

TABLE 8.2
Performance of broilers fed least-cost diets with varying levels of cassava meal

	Cassava meal in diets (percentages)			
	0	20	30	SD[1]
Chicks at end of trial (no.)[2]	141	140	137	....
Mortality[1]	4.7	5.4	7.4	....
Av. body wt/chicken (kg)				
at 7 wk	1.69	1.75	1.63	.05
at 8 wk	2.01	2.08	1.97	.08
Feed consumed/chicken (kg)				
0-7 wk	3.64	3.69	3.58	.13
0-8 wk	4.61	4.74	4.57	.18
Feed conversion[3]				
0-7 wk	2.20	2.15	2.24	.04
0-8 wk	2.34	2.33	2.36	.04

[1]Pooled standard deviation = Error mean square.
[2]Initial number of chicks per treatment: 148 with an overall average body weight of 36.3 +(-) 5g.
[3]Units of feed consumed per unit of body weight gain.
Source: Gómez *et al.*,1984.

Mattei (1984) designed a simple cassava chipping machine for production of chips for animal feed. In one version the chipper is powered by an electric motor and in the other version, by a two-stroke or four-stroke petrol engine, each with a chipping capacity of 1 tonne of cassava per hour. The drying is done on trays made of aluminium mosquito netting supported by chicken wire stretched on a strong timber frame. The dried chips are then stored in a well-ventilated area to avoid moulding. The economics of the process are favourable. A good review of simple technologies for root crop processing is provided by the United Nations Development Fund for Women (UNIFEM) publication *Root crop processing,* 1989.

TABLE 8.3

Performance of growing finishing pigs fed least-cost diets with varying levels of cassava meal[1]

	Cassava meal in diets (percentages)			
	0	20	30	SD[2]
Pigs/group (no.)	11[3]	12	12	....
Av. final wt/pig (kg)	89.9	94.7	91.1	2.20
Av. daily gain (kg)	0.77	0.82	0.78	0.02
Av. daily feed (kg)	2.55	2.77	2.54	0.06
Feed conversion	3.39	3.37	3.31	0.10

[1]Overall avg initial weight: 20.0 +(-) 1.2 kg. Experimental period: 91 days
[2]Pooled standard deviation = Error mean square.
[3]One pig was eliminated during the first two weeks of the experiment.
Source: Gómez *et al.* 1984

Some work has also been reported on the use of sweet potato as animal feed. Yang (1982) found it satisfactory for horses, mules and hogs, for lactating dairy cows when compounded with corn meal feed, and for poultry feed at 25 percent substitution for corn. Yeh *et al.* (1978) found that the digestible energy and the metabolizable energy are 91 percent of those of corn, and the nett energy is about 79 percent of that of corn as pig feed. It is not as good as corn in terms of quantity or quality of digestible protein or energy. Results in Table 8.5 indicate that sweet potato at less than 25 percent substitution will give a better result than corn alone and at about 25 percent will give a similar weight gain and efficiency as corn. Popping the chips improved starch and nitrogen digestibility as well as removing the trypsin inhibitor which might have contributed to the lowering of the feed value, but it also resulted in reducing the lysine availability. There was a significant improvement in the performance of the pig on the popped food compared to the untreated sweet potato chips. The result is also comparable to the corn diet, which improved quality and percentage lean cut, especially at the 50 percent substitution level.

TABLE 8.4

Economic assessment of poultry and pig-feeding trials[1]

	Cassava meal in diets (percentages)		
	0	20	30
Poultry trial - lot of 1 000 chicks at 7 weeks			
Chicks and fixed costs[2]	48 600	48 600	48 600
Feed cost	74 310	77 060	74 580
Interest on working capital[3]	9 218	9 425	9 239
Total expenses	132 128	135 085	132 419
Live broilers at $Col 100/kg + litter ($Col 1 220)	161 770	167 470	152 810
Net return	29 642	32 385	20 391
Return, % of expenses	22.4	24.0	15.4
Pig trial - lot of 10 pigs			
Weaned pigs at $Col 170/kg	33 830	34 170	34 000
Fixed costs, estimated	7 550	7 550	7 550
Feed cost	42 270	44 239	39 983
Interest on working capital[3]	8 365	8 596	8 153
Total expenses	92 015	94 555	89 686
Live pigs at $Col 120/kg	108 000	114 000	109 200
Net return	15 985	19 445	19 514
Return, % of expenses	17.4	20.6	21.8

[1]Results of biological evaluations have been used in these calculations. The figures are in Colombian pesos.
[2]Includes $Col 28 800 and $Col 19 800 for cost of 1 000 one-day-old chicks and fixed costs for raising them, respectively.
[3]See text for explanations.
Source: Gómez *et al.*, 1984

TABLE 8.5

Performance of fattening pigs on different proportions of corn and sweet potato chips

Corn (% in diets)	Sweet potato chips (% in diet)	Daily[1] gain (kg)	Feed/[1] gain (kg/kg)	Source
65 - 83	0	0.53	3.93	Koh *et al.*, 1960
0	56 - 72	0.37	4.79	
30 - 39	30 - 39	0.48	3.83	
63 - 81	0	0.65 ab	3.38	Tai & Lei, 1970
45 - 58	15 - 20	0.66 a	3.37	
29 - 37	29 - 37	0.62 b	3.54	
14 - 18	42 - 54	0.58 c	3.74	
0	54 - 68	0.56 d	3.81	
72 - 84	0	0.60 a	3.08 b	Yeh *et al.*, 1979
35 - 41	35 - 41	0.48 c	3.84 b	
0	69 - 81	0.44 e	4.08 a	
69 - 75	0	0.69 a	2.95 b	Yeh *et al.*, 1980
0	63 - 68	0.60 c	3.37 a	
33 - 36	33 - 36	0.66 b	3.13 b	
72 - 84	0	0.56	3.14	Lee & Lee, 1979
35 - 41	35 - 41	0.49	3.71	
0	69 - 81	0.48	3.80	

[1]Values in the same column followed by different small letters differ significantly ($P<0.01$ or $P<0.05$)
Source: Yeh, 1982.

Chen *et al.* (1979) evaluated the efficiency of gelatinization of urea-sweet potato meal (GUSP) and found that cattle on soybean meal performed better than on GUSP or urea alone, but the feed value of GUSP was better than urea. In terms of digestibility of dry matter, crude protein, crude fibre and nitrogen retention GUSP was equivalent to soybean meal. Table 8.6

TABLE 8.6

Value of sweet potato as compared to corn in various feeding trials

Animal	Substitution for corn	Comparative value	Parameter compared
Young chicks	Up to 60%	nsd	Wt gain
Cattle	100% root trimmings	80%	Wt gain
Cattle	50%	nsd	Wt gain
Dairy cattle	100%	nsd	Milk production
Dairy cattle		91%	Milk production
Dairy cattle		88%	Milk production
Dairy cattle	50%	97%	Milk production
Lambs, steers	100%	92%	Digestibility
Chicks	10 or 20%	nsd	Wt gain
Dairy cattle	100%	nsd	Milk production

Source: Yeh & Bouwkamp, 1985.

summarizes the different results obtained using sweet potato for different livestock indicating that replacement of corn by dehydrated sweet potato in the food of dairy cows could give as much milk (91-100 percent) as corn alone (Mather *et al.*, 1948; Frye *et al.*, 1948). If the orange-flesh variety is used the milk will contain more vitamin A and 30 percent more beta-carotene than milk produced using corn alone, another added advantage. Southwell *et al.* (1948) found that beef cattle fed standard rations gained 1.07 kg/day compared to 1.17 kg and 0.98 kg/day when half or all of the corn is replaced with sweet potato respectively. The feed gain ratios were 9.51, 9.31 and 9.22 respectively. Massey (1943) found in a three-year trial that substitution of corn with sweet potato led to more meat production in lamb. Lee and Young (1979) reported that chickens fed diets containing a 24 percent substitution of corn with sweet potato grew as well as on the all-corn diet, with no significant difference in carcass quality, and the egg yolk contained more vitamin A.

PRODUCTION OF SINGLE CELL PROTEIN FOR LIVESTOCK FEED

As an additional feed for livestock these root crops could serve as suitable substrates for a range of micro-organisms. Under optimal conditions from 100 kg of sweet potato containing 6.9 kg protein, a range of *Fungi imperfecti* species could produce 8.12 kg of dried mycelia, and a residue of unutilized sweet potato tissue for livestock feed, contain 31.6 kg protein i.e. the protein concentration can be increased fourfold. Analysis showed that the mycelia contain more lysine, histidine, tryptophan, methionine and tyrosine than casein (Gray and Abou-el-Seoud, 1966). Dawson *et al.* (1951) found that waste water from the starch industry could be used as a medium for the growth of the yeast *Torulopsis utilis*. With addition of ammonium hydroxide as a source of nitrogen, as much as one tonne of dried yeast containing 50 percent protein could be obtained from 100 tonnes fresh weight of sweet potato processed for starch. Similar results have been reported for cocoyam and cassava. Production of single cell protein for animal feed has reached a level of about one million tonnes per annum in the USSR, with several plants capable of producing 0.1 million tonnes annually under construction elsewhere. A pilot plant in the International Centre for Tropical Agriculture (CIAT), Colombia, using cassava as a substrate, produces a dried product with a crude protein content of 28 percent. This final dried product was incorporated into the feed of growing rats to determine the nutritive value of the unsupplemented protein. Total weight gains over a 28-day experimental period were very poor for diets based on this material. When supplemented with methionine it gave body weight gains equivalent to casein. It is not yet certain whether small-scale plants could be technically and economically feasible as at least 60 percent of the production cost is made up of the costs of the raw material.

An important factor requiring further study is the possible health risks to individuals from continued exposure to spores of the micro-organism employed in the fermentation process. Equally important is the need to examine the effects on animals to which the single cell protein is fed. Research results to date on these aspects have not revealed any disadvantageous effects (Goering, 1979).

9. Food security in developing countries

Food security has been defined by the FAO Committee on World Food Security as the "economic and physical access to food, of all people, at all times". This implies that food should be available throughout the year to sustain household energy and health, and to meet nutritional requirements. The availability of food must be coupled with the ability of every household to acquire it: it must be affordable, especially by the poor. A food security system should act as a food bank during periods of crop failure, natural disasters and external or internal hostilities.

During periods of seasonal or national food shortages the groups nutritionally at risk include poor rural and urban families, with little or no land and limited resources to meet the nutritional needs of vulnerable infants and pregnant women. To ensure access to food supplies for such groups will involve increasing their income-earning opportunities and providing adequate supplies of basic foodstuffs at prices within their reach.

Rural food security can best be met by local measures to raise farm output of locally consumed basic foodstuffs. Production of a food surplus, in response to guaranteed markets, will provide additional income for the producers, and increased food supplies which can be processed to supply convenience food products to urban areas.

The most needy families will want to increase production if promotion is focused on foods which make up a large proportion of their diet. Ideally such foods should be adapted to existing farming systems, and be capable of producing high returns to land and to labour within the constraints of unpredictable rainfall and limited inputs of capital.

Considering the situation in a number of countries, roots and tubers have many advantages as food crops for household food security, with cassava as possibly the most significant.

Cassava is already a staple food in the tropics, where it is processed in numerous ways. It has further potential for yield improvement and for conversion into a greater range of convenience products, but this will require research in food technology to design appropriate small-scale equipment for their manufacture.

Processing root crops into convenience foods will improve their being accepted by the urban population. This will lead to expanded markets and thus encourage the increased production of root crops. Use of processed foods based on local products to replace imported foodstuffs will also conserve foreign exchange.

Malnutrition has important seasonal dimensions in many countries. The hungry season is shortened when crop and variety selection extends the harvesting period, and crops are mixed to lessen the risk. Food security is improved by cultivation of drought-resistant crops, which are grown as food reserves.

During prolonged drought periods, cassava is often the only crop to survive. When properly processed as *gari,* it is safe and convenient to eat and may be stored for up to 12 months. In the South Pacific, especially in Fiji, fermented cassava products are buried in pits in the ground for months or even years and used when required. Mature cassava plants can also be left in the ground for up to three years. However, this reduces the effective land area available for subsequent crops, and also reduces the processing quality of most varieties of cassava. Root crops can provide food for consumption during preharvest periods. In Nigeria, cassava is usually the last crop in the rotation system, as it will produce reasonable yields on depleted land.

Cassava grown as a food reserve need not be harvested if domestic food supplies are plentiful. Such a crop produces a considerable biomass as roots, stems and leaves which could be incorporated into balanced feed formulas for small livestock, such as pigs. Possession of livestock provides some buffer in times of adversity, as cash raised by their sale can be used to purchase additional food.

Rosling (1987) referred to cassava as the "Cinderella of the poor" because its rapid spread in Africa led originally to an improvement in agricultural productivity and averted potential famine in some areas. This important role will decline if current agricultural productivity in Africa continues to fall. Population pressure on land means a reduced fallow period. Lack of crop rotation brings an increase in diseases and pests resulting in lower yields. It is essential to develop and improve farming systems so as to increase productivity as well as to secure and maintain soil fertility. This will enable farmers to obtain high yields from their root crops and with an appropriate government support policy will lead to the availability of root crops all year round at affordable prices. Since food energy is still a limiting nutrient in many tropical countries the effective adoption of such a policy will provide additional energy resources to improve the health of disadvantaged groups.

In many tropical countries most of the population lives in the rural areas and practises subsistence farming. Among the main crops grown for home consumption are root crops. Policy-makers often consider root crops to be cheap food meant for the poor, and direct agricultural attention toward the major cereal crops. These are relied on to increase local food production, as they did during the Green Revolution in India. This is not always feasible because the inputs required and the marketing infrastructure are not always present. Regular rainfall is a prerequisite for the successful establishment of rice, wheat and maize as well. If the rains fail, local food security may depend on drought-resistant traditional staples such as sorghum, millet and cassava.

In some parts of Asia irrigation agriculture is already established, fertilizers and pesticides are available, and conditions are appropriate for the introduction of high-yielding cereal varieties. On the other hand these countries also have their own local root crops. The yields of these crops could be considerably improved by selective breeding and increased production inputs. A production or consumption system based on only two or three food crops is extremely vulnerable, and is likely to be nutritionally unbalanced. Cultivation of roots and tubers in addition to cereals provides greater food security and a more varied and interesting diet.

Villareal (1982) has given sweet potato as an example of root crop yield potential. Experimental yields are up to 600 percent higher than farmers' yields (Table 9.1). This gap is typical for other root crops as well.

TABLE 9.1

Sweet potato yields obtained in experimental stations compared with the national average

Country	Potential yield[1]	Farmer's yield[2] (t/ha)	Yield gap	Possible improvement (%)
Tropical				
India	37	7	30	428
Philippines	35	5	30	600
Nigeria	32	13	19	146
Temperate				
Japan	35	21	14	75
Korea	43	23	20	115
USA	45	13	32	246

[1]Based on yields obtained from experimental stations.
[2]1979 national average.
Source: FAO, 1980.

Unfortunately these root crops only attain special status in times of war, calamity and famine. Yet these are staple crops that farmers are already very familiar with, offering several ideal qualities as crops for food security in the tropics. They have high tolerance for the poorer soils resulting from reduced fallowing and population pressure on the land and, in the case of cassava, tolerance for periods of drought, encountered in arid areas. Processing of root crops could also provide cottage industry employment for rural women. Some harvest of certain root crops can, if necessary, be made during the growth cycle within 70 to 90 days, though harvest should preferably not begin before a minimum period of 120 days.

In many lowland areas of Papua New Guinea, rainfall and subsistence agriculture show marked seasonal patterns. Seasonal variation in food supply is considerably reduced by the cultivation of taro, or cocoyam. As shown in Fig. 9.1 the cultivation of two species of cocoyam, *Colocasia* and *Xanthosoma*, complement one another in providing a supply of calories throughout the growing season. Banana also, as a non-seasonal producer of calories, helps to even out the food supply and ensures household food security throughout the year.

Chandra (1979) calculated the energy efficiency with which a crop uses resources to produce returns under a given cropping system in Fiji. He obtained relative values of 66 for yam, sweet potato 60, cassava 52, cocoyam 21, while other crops on the same farm including maize, rice, pulses and vegetables gave significantly lower figures, indicating that the return to energy expenditure was higher in root crops, under his experimental conditions.

As discussed earlier sweet potato can produce more calories/unit area than cereals and most other crops except sugar cane. In terms of gross monetary return/ha potato is most profitable as shown in Table 9.2 (Horton *et al.*, 1984) with US$1500/ha, followed closely by yam (US$1 469), then sweet potato, cassava and cocoyam with lower figures, owing to their relatively low prices and yields. Cereals show lower monetary returns ranging from US$366/ha for rice to US$117 for sorghum, confirming the superiority of root crops in gross return/unit of land. Cassava, yam, potato and sweet

Figure 9-1

SEASONAL PATTERN TO THE HARVESTING OF SELECTED FOOD CROPS IN PAPUA NEW GUINEA

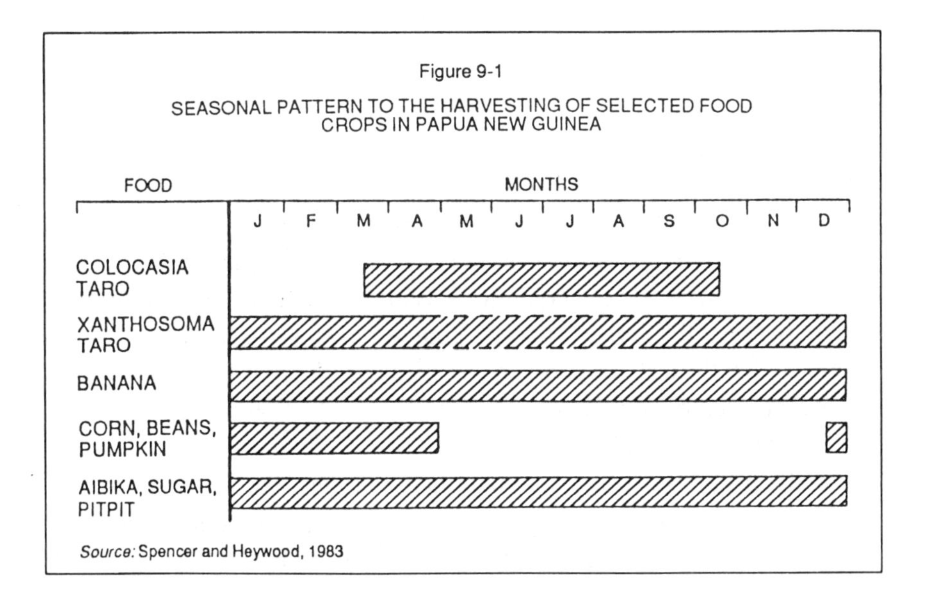

Source: Spencer and Heywood, 1983

TABLE 9.2

Average root crop and cereal grain yields, prices and gross returns/ha in developing market economies

	Yield[1] (t/ha)	Price[2] (US$/t)	Gross return[3] (US$/ha)
Potato	10.9	142	1 500
Yam	9.0	163	1 469
Sweet potato	7.1	89	629
Cassava	8.8	70	613
Cocoyam	4.2	123	514
Rice	2.2	170	366
Wheat	1.5	148	217
Maize	1.5	119	177
Sorghum	1.0	123	117

[1]Yield: average estimate for 1979/81 from FAO (1982).
[2]Price: weighted average farm-gate price corresponding to 1977, provided by FAO Basic Data Unit (unpublished).
[3]Gross return: yield multiplied by price.
Source: Horton *et al.*, 1984.

potato rank highest in the list of major food crops in terms of dry matter production per hectare (Table 9.3), potato and yam rank first and second in terms of edible energy production per hectare per day, with sweet potato ranking sixth and cassava ninth. Potato is third in the list of the most productive crops for developing market economies, in terms of edible protein per hectare per day (Horton *et al.*, 1984). Idusogie (1971) pointed out that under West African conditions, yams can provide more protein per hectare per year than maize, rice, sorghum or even soybean. Doku (1984) estimated that the use of improved varieties of root crops under conditions of good husbandry could result in an annual production of about 140 t/ha for cassava and yam, and up to 200 t/ha for sweet potato and cocoyam.

With the establishment of international institutes like the International Centre for Tropical Agriculture (CIAT), the International Institute of Tropical Agriculture (IITA), and the International Potato Center (CIP),

TABLE 9.3
Top food crops in terms of dry matter production/ha and edible energy and protein production/ha/day in developing market economies

Dry matter production		Energy production		Protein production	
crop	t/ha	crop	mj/ha/day	crop	kg/ha/day
Cassava	3.0	Potatoes	216	Cabbages	2.0
Yams	2.4	Yams	182	Dry broad beans	1.6
Potatoes	2.2	Carrots	162	Potatoes	1.4
Sweet potatoes	2.1	Maize	159	Dry peas	1.4
Rice	1.9	Cabbages	156	Eggplants	1.4
Carrots	1.7	Sweet potatoes	152	Wheat	1.3
Cabbages	1.6	Rice	151	Lentils	1.3
Bananas	1.5	Wheat	135	Tomatoes	1.2
Wheat	1.3	Cassava	121	Chickpeas	1.1
Maize	1.3	Eggplants	120	Carrots	1.0

Yield: FAO(1982) and FAO (unpublished). *Vegetative period:* FAO (1981) and Goering (1979). *Edible portion, dry matter, and food composition:* USDA (1975) and INCAP (1961)
Source: Horton *et al.*, 1984.

which are mandated to carry out research into these root crops, it is expected that governments will appreciate their economic and nutritional importance and formulate appropriate policies to encourage their production and consumption.

CONSTRAINTS TO PRODUCTION

Root crops contain about two-thirds of their weight as water. This poses two marketing problems, transportation and perishability. Cassava starts to develop a blue or brown discolouration after 24 hours, referred to as "vascular streaking". Within two days it starts to rot through the action of fungi and bacteria (Booth, 1974). If there is no central provision for processing or storage of yams and cassava these tubers have to be consumed soon after harvest within the producer's own locality. Distances to urban markets and the poor condition of rural roads often mean that the farmer has

TABLE 9.4

Comparative retail prices of some crops per 100 calories in Oceania, 1982-84

	1982	1983	1984	1982-84 average
	(Australian cents per 100 calories)			
Sweet potatoes	3.9	4.3	2.3	3.5
Taro	5.7	8.5	6.4	6.9
Rice (milled)	2.5	2.4	2.1	2.3
Wheat flour	1.7	1.8	1.9	1.8

Source: FAO, 1987b.

to accept poor prices offered by intermediaries. The alternative would involve construction of storage facilities in the localities of production.

Production of some root crops is labour intensive. Clearing the land, weeding, planting and staking in the case of yam, and harvesting, which can be single or double harvesting, all require labour. Women, who are already fully occupied with domestic duties, are also heavily involved with many of these farming activities. Yams and potatoes are regrown from previously harvested tubers. The use of small yam sets may result in reduced yields and so a proportion of the harvest, about one-fifth, is usually reserved for replanting. Assuming a yield of 12.5 t/ha and a multiplication ratio of five this can amount to as much as 2.5 t/ha reserved for replanting (Onwueme, 1978). The multiplication factor for most root crops is low when compared to cereals such as maize or guinea corn which show a multiplication ratio of 70 to 80. Some root crops such as cassava require more than a year to reach full maturity, and market handling costs are high. When root crop production is not mechanized production costs are also high. In some regions certain preferred root crops are more expensive than cereals, particularly if the latter are imported at low prices (Table 9.4).

There are a number of other production constraints involving biological problems of diseases and pests. These include the cassava mealy bug and cassava viral diseases, sweet potato viruses, root-knot nematodes and weevils. Root crops are propagated vegetatively from local planting material, which increases the risk of disease transmission.

Land holdings are small and inputs are limited in subsistence farming. In some countries the individual's land tenure is insecure as ownership of the land is vested in chiefs, landlords or the state. Pressure on agricultural land owing to population increase has resulted in much shorter fallow periods and hence less fertile soils. Traditional farming systems involve intercropping. This ensures the most effective use of family labour throughout the cropping period and provides some insurance against failure of one or more crops, but it is not conducive to the highest yields for each component.

Extension services usually limit their attention to cash crops and cereals. Subsistence farmers do not benefit from the limited research and development on root crops. The marketing and distributing infrastructure is not well developed so the farmer is reluctant to increase his acreage of root crops, and without credit facilities for fertilizers, insecticides and pesticides his crop yields are poor. The average yield of yam is only about 14 percent of the potential yield possible with optimal inputs and conditions.

The most profitable avenue for stimulating higher production of roots and tubers will be the establishment of a guaranteed market, through encouraging such industrial processing enterprises as are based on sound economic principles and are consistent with overall national development.

CONCLUSION

Root crops are essential components of the diet in many countries. In tropical Africa it has been estimated that 37 percent of the dietary energy comes from cassava. Root crops have the potential to provide more dietary energy per hectare than cereals; and some root crops, such as taro and cassava, can be grown in tropical climates all the year round, to provide increased food security. This is of particular importance during the pre-harvest period of cereal crops, when other foods are expensive or unavailable.

Attainment of food security requires that a nation should produce those products from which it will enjoy some natural and economic advantages. For many developing countries root crops offer considerable benefits and potential.

Many food-deficit countries are forced to import large quantities of grain to meet local production shortfalls. Nationally, payments for food imports are a heavy drain on foreign exchange. Increased production and consumption of domestically produced food staples such as roots, tubers and plantains will increase food supplies and broaden the food base at household and national level.

Development for the urban market of low-cost convenience foods will increase household income and stimulate increased consumption of, and demand for, these valuable foods.

Annex I

SOME RECIPES BASED ON ROOTS, TUBERS, PLANTAINS AND BANANAS

Recipes from various parts of the world are included here to illustrate the diversity of uses of a number of the root crops discussed in the preceding sections. Some of these recipes are already included in national cookery books, others are printed here for the first time as a result of a sponsored recipe competition.

The production of attractive recipe books will greatly encourage consumers to utilize more of these versatile foods.

CASSAVA RECIPES

POTATO AND SWEET POTATO RECIPES

YAM AND COCOYAM RECIPES

PLANTAIN AND BANANA RECIPES

CASSAVA ENVELOPES

Africa

Method

1. Clean, trim, steam and bone the fish.
2. Peel and dice onion, pepper and garlic. String and boil runner beans.
3. Put cassava flour in a clean dry bowl and add a pinch of salt and mix.
4. Make a well in the flour. Break eggs into the well one at a time and mix together.
5. Add some water gradually. Mix well to a dropping consistency and leave to stand for 20 minutes.
6. Heat 1 tablespoonful oil in a clean frying pan.
7. Fry fish, onions, pepper and garlic. Add salt to taste.
8. Remove from the heat when cooked, turn into a clean basin and keep covered.
9. Heat the frying pan with sufficient oil to prevent the mixture from sticking when cooking.
10. Pour mixture into the pan about two tablespoonful at a time. Make sure the mixture covers the base of the pan. Cook over a low heat until done.
11. Put a little fish mixture into the middle of each cassava cake and fold gently to look like an envelope.
12. Use a palette knife or flat spoon to remove envelopes from the pan and arrange in a dish. Repeat until the mixture is used up. Garnish with lemon slices, cooked runner beans and parsley.

Ingredients (serves four)

2 cups cassava flour
1/2 pint (285 ml) groundnut oil
1 large onion
1 sprig of parsley
1 pinch of salt
1 handful of runner beans
2 fresh or dried chili peppers
1 medium fresh fish
1 garlic clove
2 lemons (limes)
2 fresh eggs
cold water

GARI SURPRISE

Africa

Method

1. Sprinkle gari with water, mix well and leave to stand.
2. Clean, wash and cut up meat into pieces and season.
3. Season fish with cloves, garlic, cayenne pepper and salt.
4. Fry meat lightly in oil and simmer till tender. Remove and keep warm.
5. Wash, peel and slice vegetables, fry slowly for five minutes in the oil in which the meat was cooked. Remove and keep warm.
6. Lightly fry fish for five minutes.
7. Add some of the vegetables and the black-eyed beans, cook for five minutes.
8. Remove saucepan from heat. Add gari a little at a time stirring continuously. Return to heat and cook slowly for 10 minutes.
9. Cook greens in a little oil.
10. Serve gari garnished with greens and cooked meat, with remaining vegetables.

Ingredients (serves four)

2 cups fine gari
1 lb (450 g) pork or beef
2 large onions
3 large peppers
1/2 pint (285 ml) coconut oil
2 cups cooked flaked fish
1 large garden egg-plant (aubergine)
1 cup lukewarm water
1/2 lb (200 g) or cup runner beans
1/2 cup cooked black-eyed beans (cowpeas)
1 large sweet potato (orange variety)
2 cups shredded green leaves (spinach, young cabbage or sweet potato leaves)
Seasoning to taste (cayenne pepper, salt, garlic and cloves)

CASSAVA PONE (CAKE)

West Indies

Method

1. Mix the dry ingredients together.
2. Add the margarine and lard, melted together.
3. Add the eggs, vanilla essence and milk.
4. Beat well together.
5. Put into a well-greased oven pan.
6. Bake in a moderate oven for 1 1/2 hours.
7. Glaze with sugar and water and cut into squares.

Ingredients (serves six)

3 cups dry cassava flour
1 1/2 cups sugar
1/2 tsp nutmeg
1 large coconut, grated
1 1/2 cups milk
Rind of 1/2 an orange
2 oz (60 g) margarine (melted)
2 oz (60 g) lard (melted)
1/2 tsp mixed spice
1/2 tsp salt
1 tsp vanilla essence
2 eggs (beaten)

CASSAVA MEAT BALLS

Pacific Islands

Method

1. Mix together minced meat, grated cassava, onion, salt and pepper in a bowl.
2. Add one of the seasonings listed above and mix well.
3. Make into balls or small, flattened cakes (1/4" (0.5 cm) thick).
4. Roll or dip into flour.
5. Fry in hot oil until brown on each side.
6. Serve with gravy, soup, or fresh tomato sauce.

Ingredients

250 g (1/4 kg) minced beef or mutton
1 cup grated sweet cassava (raw)
2 tsp salt
1/2 tsp pepper
1 small onion, finely chopped
Flour and oil

Possible seasonings

1/4 cup chopped parsley
or 2 tsp soy sauce
or 1 tsp crushed fresh ginger
or 1 tsp mixed herbs
or 2 tsp curry powder

CASSAVA CURRY
India

Method

1. Peel the sweet cassava, cut into pieces and soak in fresh water overnight.
2. Drain cassava and squeeze dry. Boil until soft, add salt and mash well.
3. Heat oil and fry mustard, dhals, coconut and seasonings with sufficient turmeric to give a good colour to the dish.
4. Add the cooked cassava, stir well and remove from the fire. Serve hot.

Ingredients (serves four)

2 lb (1 kg) sweet cassava (soaked overnight)
1 oz (30 g) grated coconut
1 oz (25 g) Bengal gram dhal cooked
1 oz (25 g) black gram dhal cooked
Cooking oil
4 green chilies
2 in (5 cm) fresh ginger root
Salt, mustard and turmeric to taste
4 curry leaves

TAPADO
Latin America

Ingredients (serves six)

2 lb (900 g) of dried salted beef
1 lb (450 g) of cassava
2 ripe bananas
4 green plantains
1 sweet chili pepper
1 coconut
Onions and tomato to taste

Method

1. Cut the meat into small pieces and barely cover with cold water.
2. Add the chopped onions and tomatoes and season to taste.
3. Cook slowly until almost tender.
4. Peel bananas, plantain and cassava and cut into small pieces.
5. Separate the water from the coconut and put it aside. Grate the coconut flesh.
6. In a large cooking pot or oven dish place a layer of cassava, followed by a layer of banana and plantain.
7. Cover with the cooked and seasoned meat and place over this another layer of cassava and plantain.
8. Mix together the meat stock and coconut water and pour it over the layered meat and vegetables.
9. Cook until well done and serve warm.

POTATO PATCHELLI

India

Method

1. Mix together the potatoes, onion, chilies and ginger.
2. Add the thick coconut milk, and stir into the mixture to give a consistency of thick porridge.
3. Add the beaten yoghurt, salt to taste and mix well.
4. Put on the fire, bring to the boil and serve.

Note. As a variation, just before serving, you can add a few curry leaves, a little mustard seed and two dried chilies that have been fried and roughly ground.

Ingredients (serves six as a side-dish)

4 floury potatoes, boiled, drained and roughly chopped
3 or 4 fresh chilies, seeds removed, and finely chopped
1 large onion, finely chopped
6 fluid oz (180 ml) beaten yoghurt
Thick coconut milk
1 slice green ginger, crushed and chopped
Salt to taste

POTATO PANCAKES

Southeast Asia

Ingredients (serves four as a snack)

8 oz (200 g) potatoes
1 small onion, finely chopped
1 1/2 tblsp corn or maize flour
1 egg, beaten
Pinch of salt
Vegetable oil for cooking

Method

1. Grate the raw potatoes to a fine pulp.
2. Add chopped onion, cornflour, salt and beaten egg.
3. Mix well to give a dropping consistency
4. Drop in spoonfuls into a frying pan coated thinly with hot cooking oil.
5. Fry on each side (2-3 minutes) until brown.
6. Serve with hot, spiced tomato sauce.

POTATO AND CHEESE SAVOURY

Latin America

Ingredients (serves four)

3 lb (1 400 g) potatoes
4 oz (110 g) cheese (grated)
1 small white onion
3 cloves of garlic
Oil as required
1 cup of milk
50 fluid oz (1 1/2 litres) of water
Salt and pepper to taste
Bixin paste or turmeric to colour (as required)

Method

1. Chop the onion and garlic and fry in a little oil until golden.
2. Wash, peel and chop the potatoes, add to the fried onions and garlic, together with water, seasonings and bixin or turmeric, to give a good colour.
3. Boil until the potatoes are well cooked, then add the milk and cheese.
4. Return to the boil, then remove from the heat and serve, with avocado and salad.

SWEET POTATO AND BEAN POTTAGE

Africa

Ingredients (serves one, suitable for children)

2 tblsp cowpea or Bambara beans
1 medium sweet potato
1 teasp finely ground onion
1 medium size dried fish
2 tblsp palm oil
Water, as required

Method

1. Soak the beans overnight.
2. Cook the beans until they are soft.
3. Wash, peel and dice the sweet potato.
4. Add sweet potato to the beans with sufficient liquid just to cook the diced sweet potato.
5. Add ground onion, salt and dried fish.
6. Simmer until well cooked, stirring constantly.
7. Add palm oil and serve hot.

SWEET POTATO BISCUITS

Africa

Ingredients (serves four)

8 oz (200 g) sweet potato flour
3 oz (85 g) sugar
Juice and rind of 1 orange
3 oz (85 g) margarine
1 egg

Method

1. To prepare sweet potato flour, wash, peel, shred and sun dry sweet potatoes. Then pound and sieve.
2. Sieve flour into a bowl.
3. Add grated orange rind to flavour.
4. Rub fat into flour until the mixture looks like fine breadcrumbs or gari.
5. Add sugar and mix.
6. Beat up egg and add to mixture.
7. Add juice and mix to a stiff consistency that would leave the bowl clean.
8. Roll out pastry on a floured pastry board, to about 1/4" (0.5 cm) thick.
9. Cut into fancy shapes and prick with a fork.
10. Put onto a greased baking tray, glaze with water and sugar and bake.
11. Serve on a cake plate or tray.

SWEET POTATO BEVERAGE
West India

Method

1. Peel and grate the potatoes.
2. Wash and squeeze the pulp to remove free starch.
3. Squeeze lemons or limes and strain the juice.
4. Bring cloves and mace (or nutmeg) to the boil in a little water and strain off the extract.
5. Pack the potato pulp into a large stone jar.
6. Add sugar, lemon or lime juice and spice extract.
7. Add one gallon (4.5 litres) of cold water and stir until the sugar is completely dissolved.
8. Whisk in the beaten egg white, cover and allow to stand for 8 days.
9. Strain before drinking, if necessary.

Ingredients (for one gallon or 4.5 litres)

1 lb (450 g) white sweet potatoes
3 lemons or 4 large limes
3-4 lb (1 350-1 800 g) sugar
1/2 oz (15 g) cloves
1/2 oz (15 g) mace or nutmeg
1 egg white, well beaten

CANDIED SWEET POTATOES

West India

Method

1. Cook unpeeled potatoes in boiling water for 20-30 minutes.
2. Drain, peel and slice them.
3. Stir the sugar into the water and heat gently.
4. Add the butter, lime juice and spices.
5. Bring to the boil and stir until the syrup thickens.
6. Add the potato slices and cook for a further five minutes.

Note. Honey may be used in place of brown sugar.

Ingredients (serves four)

4 medium sweet potatoes
1 lb (450 g) dark brown sugar (jaggery)
1/4 pint (150 ml) (or more) water
1 oz (30 g) butter
Strained juice of one lime
Pinch of grated nutmeg
Pinch of powdered allspice

COCOYAM FRITTERS

Caribbean

Method

1. Boil unpeeled cocoyams for 10 minutes.
2. Cool, peel and grate.
3. Mix together wheat flour, baking powder, onion, parsley and seasonings.
4. Add mixed ingredients to grated cocoyam together with beaten egg.
5. Mix well to give a stiff batter.
6. Drop a spoonful at a time into very hot oil and fry until puffed and golden.
7. Drain well and serve hot, with spiced sauce or dip.

Ingredients (serves four to six as a snack)

4 cocoyams (*Xanthosoma*)
1 tblsp wheat flour
1/2 tsp baking powder
1 egg, lightly beaten
1 tsp chopped onion
1 tsp chopped parsley
Salt and cayenne pepper (to taste)
Deep fat or oil for cooking

STUFFED COCOYAM LEAVES

Pacific Islands

Method

1. Wash the young leaves and blanch briefly in boiling water.
2. Mix the cassava curry with sufficient cooked rice to absorb excess liquids.
3. Place a spoonful of this stuffing on to each leaf and roll up to form a closed parcel.
4. Pack all of the parcels closely together in a pan or baking dish previously greased with a little oil. Sprinkle with salt.
5. Add thin coconut milk or stock to cover and steam or bake until cooked (40-60 mins).

Ingredients (serves four)

12 young leaves of cocoyam (*Colocasia*)
Vegetable stock or thin coconut milk
Cassava curry mixture (see recipe for Cassava curry, p. 139)
A small quantity (1/2 cup) of cooked rice

COCOYAM AND FISH CAKES

Southeast Asia

Ingredients (serves four)

2 cups cocoyam (*Colocasia*) cooked and mashed
1 cup cooked white fish, boned and flaked
1 tsp finely chopped onion
2 tsp fresh coriander leaves, finely chopped
Salt and pepper to taste
1 egg, lightly beaten
Flour and cooking oil

Method

1. Mix together the cooked fish, cocoyam and chopped onion.
2. Add the seasonings and chopped coriander leaves and bind into a firm paste with the beaten egg.
3. Form into small balls and flatten into round cakes.
4. Coat with flour and fry in hot oil until well browned on both sides.
5. Serve with fresh tomato sauce flavoured with basil.

COCOYAM IN TOMATO SAUCE

Latin America

Method

1. Wash the cocoyams and cook them in their skins in salted water until soft.
2. Drain, peel and cut into pieces.
3. Make a sauce with the garlic, tomato and chopped onion by frying them gently in the oil.
4. Add the cooked cocoyam to the sauce together with two tablespoons of hot water.
5. Season to taste and serve hot.

Ingredients (serves five)

500 g (1/2 kg) cocoyam (*Colocasia*)
4 tblsp oil
4 tomatoes
2 cloves of garlic
1 medium onion
Salt and pepper to taste

CURRIED PLANTAIN

Caribbean

Ingredients (serves four)

6 plantains, peeled and sliced lengthways
2 tblsp vegetable oil
1 egg, lightly beaten
1 tblsp curry powder
1 tsp garam masala
2 cups coconut milk
Salt and chili pepper to taste

Method

1. Fry the ground chilies and curry powder in the oil for one minute.
2. Add the sliced plantain and fry until lightly coloured.
3. Add salt and coconut milk and cook gently over a low fire for 20 to 30 minutes.
4. Remove from the fire and stir in the garam masala and lightly beaten egg.
5. Garnish with fresh chopped coriander leaves and serve with plain boiled rice.

GREEN BANANA KORDOH

Africa

Ingredients (serves four)

- 12 green bananas
- 1/2 cup fresh groundnuts
- 3 large tomatoes, sliced
- Parsley and spring onions to garnish
- 4 medium fresh fish
- 2 onions
- Salt and fresh chili pepper to taste
- Juice of 1/2 a lemon

Method

1. Wash and peel bananas.
2. Slice and put in a covered bowl with lemon juice.
3. Roast groundnuts, peel, pound and grind to a smooth paste.
4. Scale fish, clean and remove all bones.
5. Peel onions, wash and slice, together with fresh chili pepper.
6. Boil some water in a saucepan.
7. Pound bananas and add fish, groundnuts and onions. Continue pounding until a smooth paste is formed.
8. Add salt and more black pepper to taste.
9. Remove the paste and, using wet hands, form it into small egg shapes. Put these into the saucepan to boil.
10. Boil until the paste is well cooked and the water has almost dried up.
11. Serve cold, garnished with sliced tomatoes, spring onions and parsley.

GREEN PLANTAIN PANCAKES

Latin America

Ingredients (serves four to five)

4 medium size green plantain
1/2 lb (1/4 kg) crispy pork pieces
1/2 lb (1/4 kg) pork sausages
1 small cabbage
1/2 lb (1/4 kg) tomatoes
3 hard-boiled eggs
1 sweet chili
3 tblsp cooking oil or butter
Pepper and salt to taste

Method

1. Cook the plantains until tender.
2. Grind or pound the cooked plantains with the crispy pork pieces to make a smooth paste.
3. Prepare some fresh tomato sauce, with salt and pepper to taste, and blend it with the plantain paste.
4. Chop finely the cabbage, chili, sausages and hard-boiled eggs and mix all together with a little oil.
5. Divide the plantain paste into ten pieces and flatten them to form pancakes (tortillas).
6. Spread each tortilla with a little butter, or oil, and fill centre with the chopped mixture, as prepared.
7. Fold over and wrap each tortilla in a plantain leaf.
8. Tie firmly, place in a cooking pot with a little water and steam or boil for 15 minutes.

Annex 2

MEALS FOR YOUNG CHILDREN

Adapted from *Feeding the weaning age group in the Caribbean*. Proceedings of a Technical Group Meeting. Caribbean Food and Nutrition Institute, Kingston, Jamaica, 1979.

COMPOSITE FLOUR PUDDING

Ingredients (yields 6-7 oz or 170-200 g)

2 tblsp composite flour
2 tblsp skimmed milk powder
1/4 cup fruit puree, banana or guava
1 cup water
1 tsp honey or sugar
1 tsp oil or margarine
Salt, if needed

Method

1. Mix together the flour, skimmed milk powder and salt (optional).
2. Add water and stir over a low fire for five minutes.
3. Add sugar and sieved fruit puree.
4. Reheat, add oil or margarine and mix well.
5. Sieve if necessary and allow to cool a little before serving.

Note. The composite flour used was a blend of wheat, cassava and soybean flours.

POTATO, GREEN LEAVES AND SARDINE

Method

1. Bring water to the boil and add chopped leaves and potato.
2. Cover pot and cook for 12 to 15 minutes.
3. Mash in cooking liquid.
4. To this mixture add sardine and margarine or sardine oil and salt.
5. Sieve and serve.

Variations:

1. Substitute green banana (1 1/2 medium fingers), yam, breadfruit, cocoyam or 1/2 cup cooked rice or macaroni for potato. Green bananas may need to be cooked 10 to 15 minutes more than some varieties of yam and potato, to become tender.
2. Substitute carrot, okra or pumpkin for green leaves.
3. Sausage, liver, canned mackerel or corned beef (bully beef) or 1 egg may be used instead of sardine.

Ingredients (yields 4 oz or 120 g)

3 oz (85 g) potato, sweet or Irish, peeled and diced
1 oz (30 g) fresh green leaves, finely chopped
1/2 cup water
1 oz (30 g) sardine, mashed
1 tsp margarine (optional)
Small amount of salt

YAM, CARROT AND LIVER

Using the same method described in the previous recipe, substitute yam for potato, carrot for green leaves, and liver for sardine in the same proportions.

Liver should be cooked, grated and added to hot mashed yam and vegetable. Milk may be substituted for water, if additional liquid is required when sieving.

STIRRED SWEET POTATO PUDDING

Method

1. Boil the water and add the thinly sliced sweet potato.
2. Cover and cook over low fire for 12 to 15 minutes. Remove.
3. Mash potatoes in the cooking liquid.
4. Stir in the oil or margarine, egg, sugar or honey, and milk powder to the hot mixture.
5. Sieve and serve.

Ingredients (yields 5-6 oz or 150 to 180 g)

1 small (3 oz or 85 g) sweet potato, peeled and sliced
1/4 pint water (142 ml)
1 egg
1 tblsp skimmed milk powder
1 tsp honey or dark sugar
1 tsp margarine or oil
or 2 tblsp coconut cream

GREEN BANANA OR PLANTAIN PUDDING

Ingredients (yields1 lb or 450 g)

1 1/2 medium green bananas, peeled
or 1/2 medium plantain, peeled
1 pint (500 ml) water
2 tblsp skimmed milk powder
2 tblsp honey or dark sugar
Pinch of salt

Method

1. Finely grate the plantain or bananas.
2. Beat in 1/2 cup of water until free from lumps.
3. Boil remaining water with added milk powder.
4. Add the plantain paste, stirring continuously.
5. Cook over a low fire for 15 to 20 minutes.
6. Add the oil or margarine, honey or sugar and salt to taste.
7. Cook, sieve if necessary and serve.

Nutritive values of home-made infant foods

Infant food	Moisture %	Energy (Kcal/100 g)	Protein %	Fat %	Ca mg%	Fe mg%	Na mg%	K mg%
1.Composite flour + guava	78.8	98	3.6	0.6	102	1.0	111.2	150.1
. Composite flour + banana	79.4	91	3.1	0.3	83	0.8	12.7	181.3
2.Potato, green leaves and sardine	81.0	112	4.7	4.2	65	1.2	185.5	228.8
3.Yam, carrot and liver	81.0	93	2.3	1.7	16	1.1	62.5	130.9
4.Sweet potato pudding	69.8	166	3.0	4.7	82	2.0	64.2	123.1
5.Banana pudding	89.4	45	1.8	0.4	56	0.6	140.0	151.7
Plantain pudding	80.2	84	2.5	0.4	78	1.0	84.9	136.3

Source: Feeding the weaning age group in the Caribbean. Proceedings of a Technical Group Meeting, CFNI, Kingston, Jamaica, 1979.

Annex 3

WHEATLESS BREAD

The basic starch paste

In ordinary wheat bread production, vital wheat gluten is the key ingredient that entraps fermentation gases from yeast and allows the bread to rise. In making wheatless bread, vital wheat gluten is replaced by a thick starch paste. Therefore, this is what we make first.

Start by adding 400 g of cassava or rice starch to 2 200 ml of water in a saucepan. (The starch may be replaced with cassava flour or rice flour.) The starch has a tendency to settle, so make sure you stir it up vigorously in order to keep it in suspension. Put the saucepan containing the suspended starch on the stove and start to heat it at medium to high heat with constant stirring. Do not let it burn! As you keep stirring you may see small pieces, or strands of precooked starch starting to form. This is a sign that the starch suspension will soon turn into a thick paste. Keep stirring while on medium heat. The paste will be very thick! As you continue to stir, the appearance of the paste will start to change in colour and will become more clear (or translucent). When the paste is uniformly translucent, take it off the heat and cool it with occasional stirring. It may be convenient to cool the saucepan in a sink containing cold water so that the saucepan temperature comes down quickly.

The wheatless bread batter

First measure out all the ingredients.

Yeast. Start by adding 25 g of fresh yeast in 150 ml of water and 5 g of sugar. Mash the yeast in and stir until uniform. You can also use dry yeast, 2 teaspoons in 160 ml of water with 5 g of sugar, (or follow manufacturer's instructions). This is the yeast suspension.

Sugar. 100 g or to your taste preference.

Salt. 40 g or to your taste preference.

Vegetable oil. 20 to 50 ml depending on your taste (corn oil, sunflower oil, etc.).

Basic flour ingredients. 2 000 g of either rice flour, maize flour, sorghum flour, or millet flour. You may also use cassava flour (not starch) but will have to supplement it with 50 to 70 g of high quality soy flour if you wish to bring up the protein level. You can also add 10 to 40 g of soy flour to the other flours. You can make mixtures of the various basic flour ingredients, but the total flour weight should remain in the 2 000 g range.

Baking the bread

Take all the starch paste and put it in the bowl of a mixer (use a spatula to make sure you collect all the paste). Add the flour slowly and mix with a wooden spoon or stainless steel spoon to incorporate it into the paste. This step takes some patience because the flour has a tendency to be dusty. When the flour has been moistened by the starch paste, you can put the bowl on to an electric mixer for the remaining steps. Add the sugar and salt. Mix slightly. Test the temperature to make sure that the mix is not hot, in order to be certain that the yeast is not damaged. Add the yeast suspension while stirring and you will see the batter become looser. Then, add the oil and continue to mix for five minutes. (The batter consistency will vary with the flour used.)

While the batter is being mixed, fully grease the inside of the baking pans using any vegetable oil. The best results are obtained with small rectangular pans 20 cm long, 7 cm wide and 6 cm high, with straight sides. You may experiment with different sizes and shapes of pans.

Pour the batter into the greased pans to a level of about half the height of the pan. Place the pans in a warm place (30 to 40°C) and cover with a wet cloth in order to allow the bread to rise or proof. Proofing time will vary according to the recipe. Let it proof or rise to about one centimetre below the top of the pan.

Place the pans in an oven which has been preheated to 210°C, and bake for 40 to 45 minutes. Remove the breads carefully from the pan and place in a cool area. Allow to cool for a minimum of 12 hours. (In the case of 100 percent cassava bread allow to cool 24 hours or the bread will be too sticky.) After cooling, the bread will be ready for slicing and eating.

Bibliography

Abrahamsson, L. 1978. *Food for infants and children in developing and industrialized countries.* Univ. of Uppsala, Sweden. (Ph. D. thesis)

Adewusi, S.R.A., Afolabi, O.A. & Oke, O.L. 1988. Nutritive value of cassava roots and some cereals. (Unpublished document)

Ahmed, R. 1982. Survey of glycoalkaloid content in potato tuber growing in Pakistan and environmental factors causing their synthesis and physiological investigations on feeding high glycoalkaloids to experimental animals. *6th Ann. Res. Rep.* Botany Dept, Karachi Univ., Pakistan.

Arthur, J.C. & McLemore, T.A. 1957. Effects of processing conditions on the chemical properties of canned sweet potatoes. J. *Agr. Food Chem.*, (5): 863-867.

Asenjo, C.F. & Porrata, E.I. 1956. The carotene content of green and ripe plantains. *J. Agric. Univ. Puerto Rico,* (40): 152-156.

Augustin, J., McDole, R.E., McMaster, G.M., Painter, C.G. & Sparks, W.C. 1975. Ascorbic acid content in Russet Burbank potatoes. *J. Food Sci.*, (40): 415-416.

Augustin, J., Johnson, G.K., Teitzel, C., True, R.H., Hogan, J.M. & Deutsch, R.M. 1978. Changes in nutrient composition of potatoes during home preparation. II. Vitamins. *Am. Potato J.,* (55): 653-662.

Ayangade, S.O., Oyelola, O.O. & Oke, O.L. 1982. A preliminary study of amniotic and serum thiosulphate levels in cassava eating women. *Nutri. Rep. Int.*, (26): 73-75.

Ayensu, E.S. & Coursey, D.G. 1972. Guinea yams. The botany, ethnobotany, use and possible future of yams in West Africa. *Econ. Bot.,* (26): 301-318.

Bevan, C.W.L. & Hirst, J. 1958. A convulsant alkaloid of *Dioscorea dumetorum. Chem. Ind.,* (25-Jan): 103.

Bolhuis, G.G. 1954. The toxicity of cassava roots. *Neth. J. Agric. Sci.,* (2): 176-185.

Booth, R.H. 1974. Post-harvest deterioration of tropical root crops: losses and their control. *Trop. Sci.*, (16): 49-63.

Bourdoux, P., Delange, F., Gerard, M., Mafuta, M., Hudson, A. & Ermans, M.A. 1978. Evidence that cassava ingestion increases thiocyanate formation: a possible etiologic factor in endemic goitre. *J. Clin. Endocrinol. Metab.*, (4b): 613-621.

Bourdoux, P., Seghers, P., Mafuta, M., Vanderpas, J., Vanderpas-Rivera, M., Delange, F. & Ermans, M.A. 1983. Traditional cassava detoxification process and nutrition education in Zaire. *In* Delange, F. & Akluwalia, R. eds.*Cassava toxicity and thyroid: research and public health issues*, p. 134-137. Ottawa, IDRC (IDRC 207e).

Bradbury, J.H. & Holloway, W.D. 1988. *Chemistry of tropical root crops.* Canberra, Australian Centre of International Agriculture Research.

Bruijn, G.H. 1973. The cyanogenic character of cassava. *In* Nestel, B., MacIntyre, R. eds. *Chronic cassava toxicity*, p. 43-48. Ottawa, IDRC (IDRC-10e).

Bushway, R.J., Bureau, J.L. & McGann, D.F. 1983. Alpha-chaconine and alpha-solanine content of potato peels and potato peel products. *J. Food Sci.*, (48): 84-86.

Busson, F., Jardin, C. & Wuheung, W.T. 1970. *Food composition table for use in Africa.* Rome, FAO.

Casadei, E. 1988. Nutritional and toxicological aspects of cassava. *In* Walter, R. & Quattruci, E. eds. *Nutritional and toxicological aspects of food processing*, p. 171-176. London, Taylor & Francis.

Catalano, E.A., Hasling, V.C., Dupung, H.P. & Costantin, R.J. 1977. Ipomeamarone in blemished and diseased sweet potatoes. *J. Agric. Food Chem.*, (25): 94-96.

Chandra, S. 1979. Energetics of crop production in Fiji. *Agric. Mech. Asia*, (10-3): 19-24.

Chandra, S. 1980. Root crops in Fiji, part 2: Development and future food production strategy. *Fiji Agric. J.*, (42): 11-17.

Chandra, S. 1984. *Edible aroids.* Oxford, UK, Clarendon.

Chandra, S. 1988. Tropical root crops: food for a hungry world. *Symp. Int. Soc. Root Crops*, (7): 23. Guadeloupe.

Chen, M.C., Chen, C.P. & Din, S.L. 1979. The nutritive value of sweet potato vines for cattle. V. Fresh and dehydrated sweet potato vines. *J. Agric. Assoc. China,* (107): 45-60.

Chick, H. & Slack, E.B. 1949. Distribution and nutritive value of the nitrogenous substances in potato. *Biochem. J.,* (45): 211-221.

Christiansen, J.A. 1977. *The utilization of bitter potatoes to improve food production in the high altitude of the tropics.* Cornell Univ., Ithaca, N.Y. (Ph. D. thesis)

CIP 1982. *World potato facts.* International Potato Center, Lima.

Cliff, J., Martelli, A., Molin, A. & Rosling, H. 1984. Mantakassa: an epidemic of spastic paraparesis associated with chronic cyanide intoxication in a cassava staple area of Mozambique. *WHO Bull.,* (62): 477-484.

Cock, J.H. 1985. *Cassava: new potential for a neglected crop.* Boulder (Co.), Westview, I.A.D.C.

Collins, W.W. & Walter, W.M. 1982. Potential for increasing nutritional value of sweet potatoes. *In* Villareal, R.L. & Griggs, T.D. eds. *Int. Symp. Sweet Potato. 1,* Tainan, 1982. p. 355-363. Taiwan, AVRDC.

Collis, W.R.F., Dema, I.S. & Lesi, F.E.A. 1962. Transverse survey of health and nutrition, Parkshin Division, N. Nigeria. *W. African Med. J.,* (11): 131-154.

Cooke, R.D. 1978. An enzymatic assay for total cyanide content of cassava. *J. Sci . Food Agric.,* (29): 345-352.

Cooke, R.D. & Maduagwu, E.N. 1978. The effects of simple processing on the cyanide content of cassava chips. *J. Food Technol.,* (13): 299-306.

Coursey, D.G. 1967. *Yams.* London, Longmans.

Coursey, D.G. 1968. The edible aroids. *World Crops,* (20-4): 25-30.

Coursey, D.G. 1973. Cassava as food: toxicity and technology. *In* Nestel, B. & MacIntyre, R. eds. *Chronic cassava toxicity. Proc. Interdisciplinary Workshop.* London, 29-30 Jan. 1973, p. 27-36. Ottawa, IDRC (IDRC-10e).

Coursey, D.G. 1976. The origins and domestication of yams in Africa. *In* Harlan, J.R. , Wet de J. & Stember, A.B. eds. *Origins of African plant domestication,* p. 383-408. The Hague, Mouton.

Coursey, D.G. 1983. Yams. *In* Chan, H.T. ed. *Handbook of tropical foods,* p. 555-601. New York, NY, Dekker.

Coursey, D.G. & Aidoo, A. 1966. Ascorbic acid levels in Ghanaian yams. *J. Sci. Food Agric.,* (17): 446-449.

Cox, P.A. 1980. Two Samoan technologies for breadfruit and banana preservation. *Econ. Bot.,* (34): 181-185.

Davis, N.T. & Olpin, S.E. 1979. Studies on the phytate: zinc molar contents in diets as a determinant of zinc availability to young rats. *Br. J. Nutr.,* (41): 591-603.

Dawson, P.R., Greathouse, L.H. & Gordon, W.O. 1951. Sweet potato: more than starch. In *Crops in peace and war. Yearbook of agriculture, 1950/51,* p. 195-203. Washington, USDA.

Deobald, H.J. & MacLemore, T.A. 1964. Effect of temperature, antioxidant, and oxygen on the stability of precooked dehydrated sweet potato flakes. *Food Technol.,* (18): 739-742.

Delange, F. 1983. Nutritional factors involved in the goitrogenic action of cassava. *In* Delange, F. & Ahluwalia, R. eds. *Cassava toxicity and thyroid: research and public health issues.* Ottawa, IDRC (IDRC-207e).

Devendra, C. 1977. Studies on the utilisation of cassava *(Manihot esculenta crantz)* in sheep. *MARDI Res. Bull.,* (5-2): 129-147.

Dipak, H.D. & Mukherjee, K.D. 1986. Functional properties of rapeseed protein products with varying phytic acid contents. *J. Agric. Food Chem.,* (34): 775-780.

Doku, E.V. 1969. *Cassava in Ghana.* Accra, Ghana Univ. Press.

Doku, E.V. 1984. Production potentials of major tropical root and tuber crops. *In* Terry, E.R., Doku, E.V., Arene, O.B. & Mahuneu, N.M. eds. *Tropical root crops: production and uses in Africa,* p. 19-24. Ottawa, IDRC (IDRC-221e).

Elkins, E.R. 1979. Nutrient content of raw and canned green beans, peaches and sweet potatoes. *Food Technol.,* (33-2): 66-70.

Eppendorfer, W.H., Eggum, B.O. & Bille, S.W. 1979. Nutritive value of potato crude protein as influenced by manuring and amino-acid composition. *J. Sci. Food Agric.,* (30): 361-368.

Erman, A.M., Bourdoux, P., Kinthaert, J., Lagasse, K., Luvivila, R., Mafuta, M., Thilly, C.H. & Delange F. 1983. Role of cassava in the etiology of endemic goitre and cretinism. *In* Delange, F. & Ahluwalia, R. eds. *Cassava toxicity and thyroid: research and public health issues,* p. 9-16. Ottawa, IDRC (IDRC-207e).

Essers, S. 1986. *Development of fast detoxication methods for bitter* *cassava at household level in rural N.E. Mozambique*. (Final report for Ministry of Health of Mozambique, p. 9-27) Mozambique.

FAO. 1970. Amino-acid content of food. *Nutr. Stud. No. 24*. Rome.

FAO. 1971. *Food Balance Sheets*. Rome.

FAO. 1980. *Food Balance Sheets: 1975-77 average; and per caput food supplies: 1961-65 average 1967 to 1977*. Rome.

FAO. 1984a. *Proc. Workshop on processing technologies for cassava and other tropical roots and tubers in Africa*. Abidjan, Côte d'Ivoire, 28 November - 2 December 1983. Rome.

FAO. 1984b. *Production Yearbook*. FAO Statistics Series No. 61. Rome.

FAO. 1985a. *Rep. Workshop on production and marketing constraints on roots, tubers and plantains in Africa*. Kinshasa, Zaire, 30 September - 4 October 1985. Rome.

FAO. 1985b. *Exp. cons. broadening the food base with traditional food plants*. Harare, Zimbabwe, Nov. 18-23, 1985. 108 pp. (mimeo)

FAO. 1986a. *Production Yearbook*. FAO Statistics Series No. 47. Rome.

FAO. 1986b. Role of roots, tubers and plantains in food security in sub-Saharan Africa. *Committee on World Food Security, Sess. 11*. Rome, 9-16 April 1986. Rome.

FAO. 1987a. Role of roots, tubers and plantains in food security in Latin America and the Caribbean. *Committee on World Food Security, Sess. 12*. Rome, 8-15 April 1987. Rome.

FAO. 1987b. Role of root crops in food security in the Pacific. *Committee on World Food Security, Sess. 12*. Rome, 8-15 April 1987. Rome.

FAO. 1987c. Strategies and requirements for improved production of roots, tubers and plantains. *Committee on Agriculture, Sess. 9*. Rome, 23 March - 1 April 1987. Rome.

FAO. 1988a. *Root and tuber crops, plantains and bananas in developing countries: challenges and opportunities*. Plant production and protection paper No. 87. Rome.

FAO. 1988b. *Traditional food plants*. Food and nutrition paper No. 42. Rome.

Faulks, R.M., Griffiths, W.M., White, M.A. & Tomlins, K.I. 1982. Influence of site, variety and storage on nutritional composition and cooked texture of potatoes. *J. Sci. Food Agric.*, (33): 589.

Favier, J.C., Chevassus-Agnes, S. & Gallon, G. 1971. La technologie traditionelle du manioc au

Cameroun: influence sur la valeur nutritive. *Ann. Nutr. Alim.*, (25): 1-59.

Fawcet, W. 1921. *The banana: its cultivation, distribution and commercial uses.* 2nd ed. London, Duckworth.

Fiedler, H. & Wood, J.L. 1956. Specificity studies on the B-mercaptopyruvate-cyanide transsulfuration system. *J. Bio. Chem.*, (222): 387-397.

Finglas, P.M. & Faulks, R.M. 1984. Nutritional composition of UK retail potatoes, both raw and cooked. *J. Sci. Food Agric.*, (35): 1347-1356.

Finglas, P.M. & Faulks, R.M. 1985. A new look at potatoes. *Nutr. Food Sci.*, (92): 12.

Forsyth, W.G.C. 1980. Banana and Plantains. *In* Nagy, S., & Shaw, P.E. eds. *Tropical and subtropical fruits: composition, properties and uses*, p. 258-278. Westport, Conn., AVI.

Foy, J.M. & Parratt, J.R. 1960. A note on the presence of noradrenaline and 5-hydroxytriptamine in plantain. *J. Pharm. Pharmacol.*, (12): 360-364.

Francis, B.J., Halliday, D. & Robinson, J.M. 1975. Yams as a source of edible protein. *Trop. Sci.*, (17): 103-110.

Frye, J.B., Hawkins, G.E. & Henderson, H.B. 1948. The value of winter pasture and sweet potato meal for lactating dairy cows. *J. Dairy. Sci.*, (31): 897-903.

Gebremeskel, T. & Oyewole, D.B. 1987a. *Cocoyam and the world trends of vital statistics 1965-84.* Socioecon Unit. IITA Publ. Ibadan, Nigeria.

Gebremeskel, T. & Oyewole, D.B. 1987b. *Sweet potato and the world trend of vital statistics 1965-84.* Socioecon. Unit. Ibadan, Nigeria, IITA.

Gebremeskel,T. & Oyewole, D.B. 1987c. *Yam and the world trend of vital statistics 1965-84.* Socioecon. Unit. Ibadan, Nigeria, IITA.

Gebremeskel, T. & Oyewole, D.B. 1987d. *Cassava and the world trend of vital statistics 1965-84.* Socioecon. Unit. Ibadan, Nigeria, IITA.

Goering, T.J. 1979. Tropical root crops and rural development. *World Bank Staff Working Paper No. 324.* Washington, D.C., World Bank.

Goldstein, I.J. & Hayes, C.E. 1978. The lectins: carbohydrate-binding proteins of plants and animals. *Adv. Carbohydr. Chem. Biochem.*, (35):

127-340.

Gómez, G. , Santos, J. & Valdivieso, M. 1984. Least-cost rations containing cassava meal for broilers and growing pigs. *Symp. Int. Soc. Root Crops. 6.* Lima, 21-26 Feb. 1983, p. 393-400, Lima, International Potato Center.

Goode, P.M. 1974. *Some local vegetables and fruits of Uganda.* Entebbe, Uganda, Dept of Agriculture.

Gray, W.D. & Abou-el-Seoud, M.O. 1966. Fungal protein for food and feeds II. Whole sweet potato as a substrate. *Econ. Bot.,* (20): 119-126.

Hahn, S.K., Terry, E.R., Lanschner, K., Akobundu, I.O., Okoli, C. & Lal R. 1979. Cassava improvement in Africa. *Field Crops Res.,* (2): 193-226.

Hahn, S.K. 1983. Cassava research to overcome the constraints to production and use in Africa. In Delange, F. & Akluwalia, R. eds. *Cassava toxicity and thyroid: research and public health issues,* p. 93-102. Ottawa, IDRC (IDRC-207e).

Hahn, S.K. 1984. *Tropical root crops, their improvement and utilization.* Ibadan, Nigeria, ITTA.

Hammond, A.L. 1977. Alcohol: a Brazilian answer to the energy crisis: *Sci.,* (195): 564-566.

Herrera, H. 1979. Potato protein: nutritional evaluation and utilization. Michigan State Univ. (Ph. D. thesis)

Hellendoorn, E.W., Noordhoff, M.G. & Slagman, J. 1975. Enzymatic determination of the indigestible residue (dietary fibre) content of food. *J. Sci. Food Agric.,* (26): 1461-1468.

Hentschel, H. 1969. Wertgebende Inhaltsstoffe der Kartoffel in Abhängigkeit von verschiedenen Haushaltsmassigen Zubereitungen. Mitt. 4 Vitamine und Mineralstoffe. *Qual. Plant. Mat. Veg.,* (17): 201-216.

Hoff, J.E., Junes, C.M., Wilcox, G.E. & Castro, M.D. 1971. The effect of nitrogen fertilization on the composition of the free amino-acid pool of potato tubers. *Am. Potato. J.,* (48): 390-394.

Horigone, T., Nukayama, N. & Ikeda, M. 1972. Nutritive value of sweet potato protein produced from the residual products of the sweet potato industry. *Nippon Chikasam Gakkahi,* (43): 432.

Horton, D. 1980. *The potato as a food crop for the developing world.* Lima, International Potato Center.

Horton, D., Lynam, J. & Knipscheer, H. 1984. Root crops in developing countries – an economic appraisal. In *Symp. Int. Soc. for Root Crops*. 6. Lima, 21 - 26 Feb. 1983, p. 9-39. Lima, International Potato Center.

Howlett, W.P. 1985. *Acute spastic paraplegia, Mara region, Tanzania,* Med. Assoc. Dar-es-Salaam. Sept. 1985.

Huang, P.C. 1982. Nutritive value of sweet potato. *In* Villareal, R.O. & Griggs, T.D. eds. *International Symposium on Sweet Potato. 1.* Tainan, 1982, p. 35-36. Taiwan, AVRDC.

Idusogie, E.O. 1971. The nutritive value per acre of selected food crops in Nigeria. *J. W. Afr. Sci. Assoc.*, (16): 17.

International Bank for Reconstruction and Development. 1989. Population, per capita product and growth rates. *World Bank Atls.* Washington.

Jadhav, S.J. & Salunkhe, D.K. 1975. Formation and control of chlorophyll and glycoalkaloids in tubers of *Solanum tuberosum L.* and evaluation of glyco-alkaloid toxicity. *Adv. Food Res.*, (21): 307-354.

Jones, W.O. 1959. *Manioc in Africa.* California, Stanford Univ. Press.

Kahn, E.J. 1985. The staffs of life. Boston, Mass. Little Brown.

Kay, D.E. 1973. Root Crops. *TPI Crop and Product Digest 2.* London, Tropical Products Institute.

Kay, S.J. 1985. Formulated sweet potato products. *In* J.C. Bouwkamp, ed. *Sweet potato products: a natural resource for the tropics,* p. 205-218. Boca Raton, Fl, CRC Press.

Lancaster, P.A., Ingram, J.S., Lim, M.Y. & Coursey, D.G. 1982. Traditional cassava-based foods: survey of processing techniques. *Econ. Bot.*, (36-1): 12-45.

Lawrence, G. & Walker, P. D. 1976. Pathogenesis of *E. necroticans* in Papua New Guinea. *Lancet,* (2): 125.

Lee, P.K. & Young, Y.F. 1979. Nutritive value of high protein sweet potato meal as feed ingredients for Leghorn chicks. *J. Agric. Assoc. China,* (108): 56.

Lim Han Kuo 1967. Animal feeding stuff. Part 3: compositional data on feeds and concentrates. *Malays. Agric. J.*, (46): 63-79.

Lin, S.S.M. & Chen, D.M. 1985. Sweet potato production and utilization in Asia and the Pacific. *In* Bouwkamp, J.C. ed. *Sweet potato products: a natural resource for the tropics,* p. 139-148. Boca Raton, Fl, CRC Press.

Linnemann, A.R., van Es, A. &

Hartmans, K.J. 1985. Changes in the content of L-ascorbic acid, glucose, fructose, sucrose and total glycoalkaloids in potatoes stored at 7, 16 and 28°C. *Potato Res.*, (28): 271-282.

Lopez de Romana, G., Graham, G.G. & MacLean, W.C. 1981. Prolonged consumption of potato-based diets by infants and small children. *J. Nutr.*, (111): 1430-1436.

Lopez, A., Williams, H.L. & Cooler, F.W. 1980. Essential elements in fresh and in canned sweet potatoes. *J. Food Sci.*, (45): 675-678, 681.

Lynam, J.K. & Pachico, D. 1982. *Cassava in Latin America: current status and future prospects.* Cali, CIAT.

Maga, J.A. 1980. Potato glycoalkaloids. *CRC Crit. Rev. Food Sci. Nutr.*, (12): 371-405.

Marfo, E.K. & Oke, O.L. 1988. Changes in phytate content of some tubers during cooking and fermentation. (Personal communication)

Marriott, J. & Lancaster, P.A. 1983. Bananas and plantains. *In* Chan, H.T., ed. *Handbook of tropical foods*, p. 85-143. New York, NY, Dekker.

Martin, F.W., Telek, L. & Ruberté, R.M. 1974. Yellow pigments of *Dioscorea bulbifera. J. Agr. Food Chem.*, (22-2): 335-337.

Massal, E. & Barrau, J. 1956. Banana. *In* Food Plants of the South Sea Islands, p. 15-18. *Tech. Pap. No. 94*. Noumea, South Pacific Commission.

Massey, Z.A. 1943. Sweet potato meal as livestock feed. *Georgia Agric. Expt. Sta. Bull.* No. 522.

Mattei, R. 1984. Sun drying of cassava for animal feed. A processing system for Fiji. Suva, Fiji. 43 pp. (Unpublished document)

Mather, R.E., Linkous, W.N. & Eheart, J.F. 1948. Dehydrated sweet potato as concentrate feed for dairy cattle. *J. Dairy Sci.*, (31): 569-576.

McCann, D.J. 1977. Cassava utilization in agro-industrial systems. *In* Cock, J., MacIntyre, R. & Graham, M., eds. *Symp. Int. Soc. Tropical Root Crops. 4*. CIAT, Cali, Colombia, 1-7 August 1976, p. 215-221. Ottawa, IDRC (IDRC-080e).

McCay, C.M., McCay, J.B. & Smith, O. 1975. The nutritive value of potatoes. *In* Talburt, W.F. & Smith, O. eds. *Potato Processing*. 3rd ed., p. 235-273. West Port, Conn., AVI.

McFie, J. 1967. Nutrient intakes of urban dwellers in Lagos, Nigeria. *Br. J. Nutr.*, (21): 257-268.

Meneely, G.R. & Battarblee, H.D.

1976. Sodium and potassium. *Nutr. Rev.*, (34): 225-235.

Meredith, F. & Dull, G. 1979. Amino acid levels in canned sweet potatoes and snap beans. *Food Technol.*, (33-2): 55-57.

Meuser, F. & Smolnik, H.D. 1979. Potato protein for human food use. *J. Am. Oil. Chem. Soc.*, (56): 449.

Meuser, F. & Smolnik, H.D. 1980. Processing of cassava to gari and other foodstuffs. *Starch/Starke*, (32): 116-122.

Montilla, J., Castillo, P.P. & Wiedenhofer, H. 1975. Effecto de la incorporación de harina de yuca amarga en raciones para pollos de engorde. *Agron. Trop.* (Venezuela), (25): 259-266.

Mondy, N.I. & Mueller, T.O. 1977. Potato discoloration in relation to anatomy and lipid composition. *J. Food Sci.*, (42): 14-18.

Murtin, F.W. & Ruberté, R. 1972. Yam for production of chips and french fries. *J. Agric. Univ. Puerto Rico*, (56): 228-234.

Nkamany, K. & Kayinge, K. 1982. Report of mission on spastic paralysis in the valley of rivers Lukula & Inizia in Bandundu. *Rep. No. 30*, Zaire, Dept Publ. Health, National Nutrition Planning Centre.

National Food Survey Committee 1983. Household food consumption and expenditure. *Annual report.* London, HMSO.

Nweke, F.I. 1981. Consumption patterns and their implications for research and production in tropical Africa. *In* Terry, E.R., Oduro, K.A. & Caveness, F. eds. Tropical root crops; research strategies for the 1980s. *Tri. root crops sym. 1.* 8-12 Sept. 1980. Proc. Int. soc. trop. root crops, Africa branch. Ibadan, Nigeria. p. 88-94. Ottawa, IDRC (IDRC-163e).

O'Dell, B.L. & Savage, J.E. 1960. Effect of phytic acid on zinc availability. *Proc. Soc. Exp. Biol. Med.*, (103): 304-309.

Ojo, G.O. 1969. Plantain meals and serum 5-hydroxytryptamine in healthy Nigerians. *W. Afr. Med. J.*, (18): 174.

Oke, O.L. 1966. Chemical studies on some Nigerian foodstuffs: gari. *Nature*, (212): 1055-1056.

Oke, O.L. 1967. The present state of nutrition in Nigeria. *World Rev. Nutr. Diet*, (8): 25-61.

Oke, O.L. 1968. Cassava as food in Nigeria. *WorldRev.Nutr.Diet*, (9): 227-250.

Oke, O.L. 1969. The role of hydrocyanic acid in nutrition. *World Rev. Nutr. Diet*, (11): 170-198.

Oke, O.L. 1972. Yam: a valuable source

of food and drugs. *World Rev. Nutr. Diet,* (15): 156-184.

Oke, O.L. 1973. The mode of cyanide detoxication. *In* Nestel, B. & McIntyre, E.R. *Chronic cassava toxicity.* Interdisc. Workshop Proc. London, 29-30 Jan. 1973, p. 97-104. Ottawa, IDRC (IDRC-010e).

Oke, O.L. 1984. Processing and detoxification of cassava. In *Symp. Int. Soc. Root Crops. 6.* Lima, 21-26 February 1983. p. 329-336. Lima. International Potato Center.

Oke, O.L. 1986. Cyanide detoxification mechanism by palm oil. *Proc. Am. Soc. Expt. Biol. Med.,* (5): 10.

Oke, O.L. & Ojofeitimi, E.O. 1980. Cocoyam - a neglected tuber. *World Rev. Nutr. Diet,* (34): 133-143.

Okigbo, B.N. 1978. Cropping systems and related research in Africa. *AAASA Occasional Publ.* Ser OT-1. 81 pp.

Omole, A., Adwusi, S.R.A., Adeyemo, A. & Ohe, O.L. 1978. The nutritive value of tropical fruits and root crops. *Nutr. Rep. Int.,* (17): 575-580.

Oñate, L.U., Cabotaje, E. & Barba, C.V. 1976. *Human nutrition in South East Asia.* 2nd ed. Quezon City, Philippines, Agrix, 230 pp.

Onayemi, O. & Potter, N.N. 1974. Preparation and storage properties of drum dried white yam *(Dioscorea rotundata Poir)* flakes. *J. Food Sci.,* (39): 559-562.

Onwueme, I.C. 1978. *The tropical tuber crops: yams, cassava, sweet potato and cocoyams.* Chichester, U.K. Wiley. 234 pp.

Onwueme, I.C. 1984. The place of the edible aroids in tropical farming systems. *In* Chandra, S. ed. *Edible aroids,* p. 136-139. Oxford, UK, Clarendon.

Osuntokun, B.O. 1968. An ataxic neuropathy in Nigeria: a clinical biochemical and electro-physiological study. *Brain,* (91): 215-248.

Osuntokun, B.O. 1981. Cassava diet, chronic cyanide intoxication and neuropathy in the Nigerian Africans. *World Rev. Nutr. Diet.,* (36): 141-173.

Page, E. & Hanning, F.M. 1963. Vitamin B_6 and niacin in potatoes. *J. Am. Diet. Assoc.,* (42): 42-45.

Palmer, J.K. 1982. Carbohydrates in sweet potato. *In* Villareal, R.L. & Griggs, T.D. eds. *Int. Symp. Sweet Potato. 1.* Tainan, 1982. p. 135-140. Taiwan, AVRDC.

Panalaks, T. & Murray, T.K. 1970. The effect of processing on the content of carotene isomers in vegetables and peaches. *Can. Inst.*

Food Technol. J., (3): 145-151.

Payne, P.R. 1969. Effect of quantity and quality of protein on the protein values of diets, *Voeding*, (30): 182-191.

Pena, R.S. de la & Pardales, J.R. 1984. Evidence of proteolytic enzyme activity in taro, *Colocasia esculenta*. *Symp. Int. Soc. Root Crops. 6*. Lima, 21-26 Feb. 1983, p. 157-159. Lima, International Potato Center.

Philbrick, D.J., Hill, D.C. & Alexander, J.C. 1977. Physiological and biochemical changes associated with linamarin and administration to rats. *Toxicol. Asql. Pharmacol.*, (42): 539.

Pineda, M.J. & Rubio, R.R. 1972. Un concepto nuevo en el levante de novillas para ganaderia de leche. *Rev. ICA*, (7): 405-413.

Platt, B.S. 1962. Table of representative values of foods commonly used in tropical countries. *Spec. Rep. Ser. Med. Res. Coun. No. 302*. London, HMSO.

Plucknett, D.L. 1984. Presidential address: tropical root crops in the eighties. *Symp. Int. Soc. Root Crops. 6*. Lima, 21-26 Feb, 1983, p. 3-8. Lima, International Potato Center.

Plucknett, D.L., Pena, R.S. de la & Obrero, F. 1970. Taro *(Colocasia esculenta)*. *Field Crop. Abstr.*, (23): 413-426.

Purcell, A.E. & Walter, W.M. 1982. Stability of amino acids during cooking and processing of sweet potatoes. *J. Agric. Food Chem.*, (30): 443-444.

Purseglove, J.W. 1968. *Tropical Crops: Dicotyledons. 2*. London, Longman.

Purseglove, J.W. 1972. *Tropical Crops: Monocotyledons*. London, Longman.

Rasper, V. 1969. Investigations on starches from major starch crops grown in Ghana. II. Swelling and solubility patterns: amyloclastic susceptibility. *J. Sci. Food Agric.*, (20): 642-646.

Rasper, V. 1971. Investigations on starches from major starch crops grown in Ghana. III. Particle size and particle size distribution. *J. Sci. Food Agric.*, (22): 572-580.

Rasper, V. & Coursey, D.G. 1967. Properties of starches of some West African yams. *J. Sci. Foods Agric.*, (18): 940-944.

Roine, P., Wickmann, K. & Vihavainen, L. 1955. The content and stability of ascorbic acid in different potato varieties in Finland. *Suom. Maataloust. Seur. Julk.*, (83): 71-87.

Rose, M.S. & Cooper, L.F. 1907. The biological efficiency of potato nitrogen. *J. Biol. Chem.*, (30): 201.

Rosling, H. 1987. *Cassava toxicity and food security.* Uppsala, Sweden, Tryck Kontakt. 40 pp.

Roy-Choudhuri, R.N. 1963. Nutritive value of poor Indian diets based on potato. *Food Sci.*, (12): 258.

Sakamoto, S. & Bouwkamp, J.C. 1985. Industrial products from sweet potatoes. *In* Bouwkamp, J.C. ed. *Sweet potato products: a natural resource for the tropics,* p. 219-259. Boca Raton, F.L. CRC Press.

Salaman, R.N. 1949. *The history and social influence of the potato.* Cambridge, UK, Cambridge University Press.

Sanint, L.R., Rivas, L., Duque, M.C. & Sere, C. 1985. Análisis de los patrones de consumo de alimentos en Colombia a partir de la encuesta de hogares DANE/DRI de 1981. *Rev. Plan. Des.*, (17-3): 39-68.

Satin, M. 1988. Bread without wheat. *New Scientist.* 28 April 1988.

Scott, G. 1985. *Mercados, Mitos e Intermediarios.* Lima, Universidad del Pacifico.

Shaper, A.G. 1967. Plantain diets, serotonin and endomyocardial fibrosis. *Am. Heart J.*, (73): 432.

Simmonds, N.W. 1962. *The evolution of bananas.* London, Longman.

Simmonds, N.W. 1966. *Bananas,* 2nd ed. London, Longman.

Simmonds, N.W. 1976. Banana. *In* Simmonds, N.W. *Evolution of crop plants,* p. 211-215. London, Longman.

Singh, M. & Krikorian, A.D. 1982. Inhibition of trypsin activity in vitro by phytate. *J. Agr. Food Chem.*, (30): 799-800.

Smith, A.D.M. & Duckett, S. 1965. Cyanide, vitamin B_{12} experimental demyelination and tobacco ambliopia, *Br. J. Exp. Path.*, (46): 615-622.

Southwell, B.L. & Black, W.H. 1948. Dehydrated sweet potato for fattening steers. *Georgia Agric. Expt. Sta. Bull 45.*

Spencer, T. & Heywood, P. 1983. Seasonality, subsistence agriculture and nutrition in a lowlands community of Papua New Guinea. *Ecol. Food Nutr.*, (13): 221-229.

Stanton, W.R. & Wallbridge, A. J. 1969. Fermented food processes. *Process biochem.*, (4): 45-51.

Steele, W.J.V. & Sammy, G.M. 1976. The processing potentials of yams. I. Laboratory studies on lye peeling of yams. *J. Agric. Univ. Puerto Rico,* (60): 207-214.

Streghtoff, F., Munsell, H.E., Ben-Dor, B., Orr, M.L., Caillean, R., Leonard, M.H., Ezekiel, S.R. &

Roch, K.G. 1946. Effect of large-scale methods of preparation on vitamin content of food. I. Potatoes *J. Am. Diet Assoc.*, (22): 117-127.

Swaminathan, K. & Gangwar, B.M.L. 1961. Cooking losses of vitamin C in Indian potato varieties. *Indian Potato J.*, (3): 86-91.

Sweeney, J.P., Hepner, P.A. & Libeck, S.Y. 1969. Organic acids, amino-acid and ascorbic acid content of potatoes as affected by storage conditions *Am. Potato J.*, (46): 463-469.

Tamate, J. & Bradbury, J.H. 1985. Determination of sugars in tropical root crops using CN.m.r. spectroscopy: comparison with the H.p.l.c. method. *J. Sci. Food Agric.*, (36): 1291-1302.

Taylor, J.M. 1982. Commercial production of sweet potatoes for flour and feeds. *In* Villareal, R.L. & Griggs, T.D. eds. *Int. Symp. Sweet Potato. 1.* Tainan, 1982, p. 393-404. Taiwan, AVRDC.

Treadway, R.H., Heisler, E.G., Whittenberger, R.T., Highland, S.M.E. & Getchell, Y.G. 1955. Natural dehydration of cull potatoes by alternate freezing and thawing. *Am. Potato J.*, (32): 293-303.

True, R.H., Hogan, J.M., Augustin, J., Johnson, S.R., Teitzel, C. & Show, R.L. 1978. Mineral composition of freshly harvested potatoes. *Am. Potato. J.*, (55): 511-519.

True, R.H., Hogan, J.M., Augustin, J., Johnson, S.R., Teitzel, C. & Show, R.L. 1979. Changes in the nutrient composition of potatoes during home preparation III. Minerals. *Am. Potato J.*, (56): 339-350.

United Nations 1975. *Poverty, unemployment and development policy; a case study of selected issues with references to Kerala.* New York, N.Y. United Nations. 235 pp.

United Nations Development Fund for Women. 1989. *Root crop processing.* Food Cycle Technology Source Book No. 5.

Uritani, I. 1967. Abnormal substances produced in fungus-contaminated foodstuffs. *J. Assoc. Offic. Agric. Chem.*, (50): 105-114.

Villareal, R.L. 1970. The vegetable industry's answer to the protein gap among low-salaried earners. *Sugar News* (Manila), (46): 482-488.

Villareal, R.L. 1982. Sweet potato in the tropics: progress and problems. *In* Villareal, R.L. & Griggs, T.D. eds. *Int. Symp. Sweet Potato. 1.* Tainan, 1982, p. 3-15. Taiwan,

AVRDC.

Walter, W.M. & Catignani, G.L. 1981. Biological quality and composition of sweet potato protein fractions. *J. Agr. Food Chem.,* (29): 797-799.

Walter, W.M., Catignani, G.L., Yow, L.L. & Porter, D.H. 1983. Protein nutritional value of sweet potato flour. *J. Agr. Food Chem.,* (31): 947-949.

Whitby, P. 1969. *Report on review of information concerning food consumption in Ghana.* Rome, FAO. 68 pp.

Wilson, J.E. 1984. Cocoyam. *In* Goldsworthy, P.R. & Fisher, N.M. eds. *The physiology of tropical field crops,* p. 589-605. Chichester, UK, Wiley.

Wilson, L.A. 1977. Root crops. *In* Alvim, P. de T. & Kozlowski, T.T. eds. *Ecophysiology of tropical crops,* p. 187-236. New York, NY, Academic Press.

Wokes, F. & Picard, C. W. 1955. The role of vitamin B_{12} in human nutrition. *Am. J. Clin. Nutr.,* (3): 383-390.

Wood, J.L. & Cooley, S.L. 1956. Detoxication of cyanide by cystine. *J. Biol. Chem.,* (218): 449-457.

Wood, F.A. & Young, D.A. 1974. TGA in potatoes. *Canada Dept. of Agric. Publ. No. 1533.*

Woolfe, J.A. 1987. *The potato in the human diet.* Cambridge, UK, Cambridge Univ. Press.

World Health Organization. 1985. Energy and Protein Requirements. Report. *Tech. Rep. Ser. No. 724.*

Yamaguchi, M., Perdue, J.W. & Mac Gillivrary, J.H. 1960. Nutrient composition of white rose potatoes during growth and after storage. *Am. Potato J.,* (37): 73.

Yamaguchi, Y. , Mahungu, N.M. & Hahn, S.K. 1981. *Effect of processing of cassava storage on cyanide content.* Ibadan, Nigeria, IITA. (Report)

Yang, T.H. 1982. Sweet potato as a supplemental staple food. *In* Villareal, R.L. & Griggs, T.D. eds. *Int. Symp. Sweet Potato. 1.* Tainan, 1982, p. 31-36. Taiwan, AVRDC.

Yeh, T.P., Wung, S.C., Lin, K. & Kuo, C.G. 1978. Studies on different methods of processing some local feed materials to enhance their nutritive value for swine. Anim. Ind. Res. Inst. Taiwan Sugar Corp. *Ann. Res. Report:* p. 25. (Original Chinese with English summary)

Yeh, T.P. 1982. Utilization of sweet potatoes for animal feed and industrial uses: potential and problems. *In* Villareal, R.L. & Griggs, T.D. eds. *Int. Symp. Sweet Potato. 1.* Tainan, 1982, p. 385-392. Taiwan, AVRDC.

Yen, D.E. 1978. The storage of cassava in Polynesia. Islands. *Cassava Newsletter,* (3): 9-11.

Yet, T.P. & Bouwkamp, J.C. 1985. Roots and vines as animal feed. *In* Bouwkamp, J.C., ed. *Sweet potato products: a natural resource for the tropics.* CRC Press.